Nutritional Marine Life

Nutritional Marine Life

RAMASAMY SANTHANAM

CRC Press
Taylor & Francis Group
Boca Raton London New York

CRC Press is an imprint of the
Taylor & Francis Group, an **informa** business

CRC Press
Taylor & Francis Group
6000 Broken Sound Parkway NW, Suite 300
Boca Raton, FL 33487-2742

First issued in paperback 2019

ISBN-13: 978-1-4822-6205-6 (hbk)
ISBN-13: 978-1-138-38224-4 (pbk)

Visit the Taylor & Francis Web site at
http://www.taylorandfrancis.com

and the CRC Press Web site at
http://www.crcpress.com

CONTENTS

Preface..xv
About the Author xvii

Chapter 1: Introduction..1
Nutritional Composition of Marine Foods: Fish, Crustaceans, and Mollusks.....1
    Proteins..1
    Fat and Fatty Acids..2
    Minerals and Trace Elements..2
    Vitamins 3
Seaweeds...3
    Protein..3
    Amino Acids..3
    Lipids and Fiber 4
    Minerals and Trace Elements..4
    Vitamins 4
    Pharmaceutical Compounds..4

Chapter 2: Phytoplankton 7
Nannochloropsis oculata (Droop) D.J. Hibberd, 1981 7
Nannochloropsis gaditana L.M. Lubián, 1982................................8
Phaeodactylum tricornutum Bohlin, 1897 9

Chapter 3: Seaweeds and Marsh Plants 11
Green Algae 11
    Codium tomentosum Stackhouse, 1797................................ 11
    Codium iyengarii Børgesen 12
    Caulerpa lentillifera J. Agardh 12
    Caulerpa racemosa (Forsskål) J. Agardh, 1873 13
    Halimeda tuna (J. Ellis & Solander) J.V. Lamouroux, 1816........14
    Rhizoclonium implexum (Dillwyn) Kützing, 1845........................15
    Cladophora glomerata (L.) Kutzin 1843 16
    Ulva clathrata (Roth) C. Agardh 1811 = *Enteromorpha clathrata* 17
    Ulva pertusa Kjellman, 1897..17
    Ulva intestinalis Linnaeus..18
    Ulva faciata Delile 1813..19
    Ulva lactuca Linnaeus 20
    Ulva reticulata Forsskål, 1775 21
Brown Algae 23
    Ascophyllum nodosum (L.) Le Jolis...................................... 23

Himanthalia elongata (Linnaeus) S.F. Gray 24
Hormophysa cuneiformis (Gmelin) Silva 25
Sargassum tenerrimum J. Agardh 1848 26
Sargassum variegatum 26
Sargassum wightii Greville 27
Sargassum polycystum C. Agardh 28
Turbinaria conoides (J. Agardh) Kutzing 1860 29
Calpomenia sinuosa (Mertens ex Roth) Derbès & Solier, 1851 30
Chnoospora minima (Hering) Papenfuss, 1956 31
Dictyota dichotoma var. *intricata* (C. Agardh) Greville, 1830 32
Dictyota indica Sonder ex Kützing 33
Padina australis Hauck 34
Padina durvillei Bory Saint-Vincent 34
Padina fernandeziana Skottsberg & Levring, 1941 35
Padina gymnospora (Kutzing) Sonder 36
Padina pavonica (Linnaeus) J.V. Lamouroux 37
Padina vickersiae Hoyt in Britton & Millspaugh 1920 38
Padina tetrastromatica Hauck 38
Stypopodium schimperi (Kützing) M. Verlaque & Boudouresque 1991 39
Eisenia arborea Aresch 1876 40
Laminaria digitata (Hudson) J.V. Lamouroux 41
Saccharina latissima (Linnaeus) J.V. Lamouroux (= *Laminaria saccharina*) 42
Undaria pinnatifida (Harvey) Suringar, 1873 43
Red Algae 45
Chondrus crispus Stackh 45
Kappaphycus cottonii (Weber-van Bosse) Doty (= *Eucheuma cottonii*) 46
Eucheuma denticulatum (var. yellow) (Burman) Collins et Hervey
    (= *Eucheuma spinosum*) 47
Solieria robusta (Greville) Kylin 47
Spyridia filamentosa (Wulfen) Harvey 48
Hypnea pannosa J. Agardh, 1847 49
Hypnea musciformis (Wulfen) J.V. Lamouroux 1813 50
Hypnea japonica Tanaka 1941 50
Hypnea charoides Lamouroux 51
Kappaphycus alvarezii (Doty) Doty nov. comb. (= *Eucheuma alvarezii*) 52
Kappaphycus striatum (Schmitz) Doty nov. comb. (= *Eucheuma striatum*) 54
Acanthopora spicifera (Vahl) Borgesen 55
Dasya rigidula (Kützing) Ardissone 56
Gelidium pusillum (Stackhouse) Le Jolis 57
Laurencia papillosa (C. Agardh) Greville, 1830 58
Pterocladia capillacea (Gmelin) Santelices & Hommersand 59
Gracilaria cornea J. Agardh 60
Gracilaria changgi B.M. Xia & I.A. Abbott 60
Gracilaria compressa Grev. 61

Gracilaria manilaensis Yamamoto & Trono 62
Gracilaria salicornia (C. Agardh) Dawson 1954 63
Gracilaria verrucosa (Hudson) Paperfuss 64
Palmaria palmata (Linnaeus) Weber & Mohr 65
Jania rubens (Linnaeus) Lamouroux 66
Porphyra tenera Kjellman, 1897 67
Porphyra haitanensis T. J. Chang et B. F. Cheng 67
Porphyra umbilicalis Kützing 68
Halophyte 69
Salicornia bigelovii Torr. 69

Chapter 4: Jellyfish 71
Rhopilema esculentum Kishinouye, 1891 71
Stomolophus meleagris (Agassiz, 1860) 72

Chapter 5: Crustaceans 75
Shrimps 75
Pandalus jordani Rathbun, 1902 75
Pandalus borealis (Krøyer, 1838) 75
Fenneropenaeus penicillatus (Alcock, 1905) 76
Parapenaeus longirostris (Lucas, 1847) 77
Plesionika martia (Milne-Edwards, 1883) 77
Litopenaeus vannamei (Boone, 1931) 78
Penaeus monodon Fabricius, 1798 79
Litopenaeus stylirostris Stimpson, 1874 80
Acetes japonicus Kishinouye, 1905 80
Metapenaeus affinis (Milne-Edwards, 1837) 81
Penaeus indicus (H. Milne Edwards, 1837) 82
Crabs 83
Portunus pelagicus (Linnaeus, 1758) 83
Portunus sanguinolentus (Herbst, 1783) 84
Scylla serrata Forsskål, 1775 85
Callinectes sapidus Rathbun, 1896 85
Cancer magister (Dana, 1852) 86
Paralithodes camtschatica (Tilesius, 1815) 87
Chionoecetes opilio (O. Fabricius, 1788) 88
Geryon quinquedens Smith, 1879 88
Pleuroncodes planipes Stimpson, 1860 89
Chionoecetes bairdi Rathbun, 1924 89
Lobsters 90
Panulirus interruptus (J.W. Randall, 1840) 90
Homarus gammarus Linnaeus, 1758 91
Homarus americanus H. Milne-Edwards, 1837 92
Linuparus somniosus Berry & George, 1972 93

Chapter 6: Mollusks 95
  Oysters 95
    Crassostrea gryphoides (Schlotheim 1813) 95
    Crassostrea rivularis (Gould 1861) 96
    Crassostrea madrasensis (Preston) 96
    Crassostrea gigas (Thunberg, 1793) 97
    Saccostrea cucullata (Born, 1778) 97
  Mussels 98
    Mytilus edulis Linnaeus 1758 98
    Mytilus galloprovincialis Lamarck, 1819 99
    Modiolus (Modiolus) modiolus (Linnaeus, 1758) 100
    Lithophaga lithophaga (Linnaeus, 1758) 100
  Clams 101
    Ensis directus Conrad, 1843 101
    Arctica islandica Linnaeus, 1767 102
    Donax trunculus Linnaeus, 1758 102
    Meretrix meretrix (Linnaeus, 1758) 103
    Meretrix lusoria Roeding, 1798 104
    Venerupis philippinarum (A. Adams & Reeve, 1850) 104
    Saxidomus giganteus (Deshayes, 1839) 105
    Protothaca staminea (Conrad, 1837) 105
    Venerupis decussata (Linnaeus, 1758) 106
    Mercenaria mercenaria (Linnaeus, 1758) 107
    Mactra sachalinensis Schrenck, 1862 107
    Mya arenaria Linnaeus, 1758 108
    Panopea generosa Gould, 1850 109
  Scallops 110
    Chlamys farreri (Jones & Preston, 1904) 110
    Patinopecten yessoensis Jay, 1857 110
    Pecten albicans (Schröter, 1802) 111
  Conches 112
    Strombus gigas Linnaeus, 1758 112
    Babylonia spirata (Linnaeus, C., 1758) 113
    Bullacta exarata (Philippi, 1849) 114
  Abalones 115
    Haliotis spadicea Donovan, 1808 115
    Haliotis cracherodii Leach, 1814 115
    Haliotis tuberculata Linnaeus, 1758 116
    Haliotis rufescens Swainson, 1822 116
    Haliotis rubra W.E. Leach, 1814 117
    Haliotis gigantea Gmelin, 1791 117
    Haliotis discus hannai Ino, 1953 118
  Limpet 119
    Cellana exarata (Reeve, 1854) 119

Squids 120
   Todarodes pacificus (Steenstrup, 1880) 120
   Ommastrephes bartramii (Lesueur, 1821) 121
   Illex argentines (Castellanos, 1960) 121
   Uroteuthis chinensis (Gray, 1849) 122
Cuttlefish 123
   Sepia esculenta Hoyle, 1885 123
   Sepiella japonica Sasaki, 1929 123
Octopus 124
   Enteroctopus dofleini (Wülker, 1910) 124
   Octopus ocellatus Gray, 1849 125

Chapter 7: Echinoderms 127
Starfish 127
   Asterias amurensis Lütken, 1871 127
Sea Urchins 128
   Tripneustes gratilla (Linnaeus, 1758) 128
   Strongylocentrotus droebachiensis (O.F. Müller, 1776) 129
   Strongylocentrotus franciscanus (A. Agassiz, 1863) 130
Sea Cucumbers 131
   Acaudina molpadioides (Semper, 1867) 131
   Thelenota ananas Jaeger, 1833 131
   Apostichopus japonicus (Selenka, 1867) 132
   Holothuria (Metriatyla) scabra Jaeger, 1833 133
   Holothuria (Microthele) nobilis (Selenka, 1867) 133
   Holothuria (Thymiosycia) impatiens (Forskål, 1775) 134
   Holothuria pardalis Selenka, 1867 (= *Holothuria (Lessonothuria) insignis*) 134
   Holothuria (Lessonothuria) multipilula Liao, 1975 135
   Actinopyga echinites (Jaeger, 1833) 135

Chapter 8: Prochordate 137
   Halocynthia roretz (VonDrasche, 1884) 137
   Herdmania pallida (Heller, 1878) 138

Chapter 9: Fish 141
Teleosts 141
   Chirocentrus dorab (Forsskål, 1775) 141
   Dussumieria acuta Valenciennes, 1847 142
   Ilisha melanostoma (Schneider, 1801) 142
   Pellona ditchela Valenciennes, 1847 142
   Sardinella longiceps Valenciennes, 1847 143
   Sardinella fimbriata (Valenciennes, 1847) 143
   Ilisha elongata (Anonymous [Bennett], 1830) 144
   Tenualosa macrura (Bleeker, 1852) (= *Hilsa (Clupea) macrura*) 145

Contents

Dussumieria hasselti Bleeker, 1849 146

Clupea harengus Linnaeus, 1758 146

Ethmalosa fimbriata (Bowdich, 1825) 147

Anadontosoma chacunda (Hamilton, 1822) 148

Stolephorus commersonii Lacepède, 1803 149

Cottoperca gobio (Gunther, 1861) 149

Brama brama (Bonnaterre, 1788) 150

Caesio erythrogaster (Bloch, 1791) 151

Trachinotus blochii (Lacepède, 1801) 151

Parona signata (Jenyns, 1841) 152

Seriola lalandi Valenciennes, 1833 153

Parastromateus niger (Bloch, 1795) 153

Selaroides leptolepis (Cuvier, 1833) 154

Megalapsis cordyla (Linnaeus, 1758) 155

Carangoides malabaricus (Bloch & Schneider, 1801) 156

Carangoides orthogrammus (Jordan & Gilbert, 1882) 157

Selar mate (G. Cuvier, 1833) 157

Chorinemus lysan (Forsskål, 1775) 158

Caranx djeddaba (Forsskål, 1775) 159

Decapterus russelli (Rüppell, 1830) 160

Seriola quinqueradiata Temminck Schlegel, 1845 160

Seriolella punctata (Forster, 1801) 161

Nemadactylus bergi (Norman, 1937) 162

Drepane punctata (Linnaeus, 1758) 162

Thalassoma fuscum (Lacepède, 1801) (= *Thalassomatrilobatum*) 163

Cheilinus undulatus Rüppell, 1835 164

Lates calcarifer (Bloch, 1790) 164

Leiognathus dussumieri (Valenciennes, 1835) (= *Karalla dussumieri*) 165

Gazza achlamys Jordan Starks, 1917 166

Leiognathus equulus (Forsskål, 1775) 167

Lethrinus lentjan (Lacepède, 1802) 168

Lutjanus quinquelineatus (Bloch, 1790) 168

Lutjanus lutjanus (Bloch, 1790) 169

Lutjanus decussatus (Cuvier, 1828) 169

Lutjanus argentimaculatus (Forsskål, 1775) 170

Lutjanus gibbus (Forsskål, 1775) 171

Lutjanus russelli (Bleeker, 1849) 172

Pristipomoides typus Bleeker, 1852 173

Lutjanus johnii (Bloch, 1792) 173

Lutjanus malabaricus (Bloch & Schneider, 1801) 174

Lutjanus bohar (Forsskål, 1775) 175

Monodactylus argenteus (Linnaeus, 1758) 176

Morone saxatilis (Walbaum, 1792) 177

Dicentrarchus labrax (Linnaeus, 1758) 178

Parupeneus bifasciatus (Lacepède, 1801) 178

x

Nemipterus bleekeri (Day, 1875) (= *Nemipterus bipunctatus*) 179
Nemipterus japonicus (Bloch, 1791) 180
Scolopsis bimaculatus Rüppell, 1828 180
Dissostichus eleginoides Smitt, 1898 181
Patagonotothen ramsayi (Regan, 1913) 182
Perca fluviatilis Linnaeus, 1758 183
Pseudopercis semifasciata (Cuvier, 1829) 183
Plectorhinchus pictus (Tortonese, 1936) 184
Pomadasys hasta (Bloch, 1790) 185
Eleutheronema tetradactylum (Shaw, 1804) 185
Pomatomus saltatrix (Linnaeus, 1766) 186
Sciaenops ocellatus (Linnaeus, 1766) 188
Atractoscion nobilis (Ayres, 1860) (= *Cynoscion nobilis*) 188
Cynoscion nebulosus (Cuvier, 1830) 189
Cynoscion regalis (Bloch & Schneider, 1801) 189
Leiostomus xanthurus Lacepède, 1802 190
Johnius dussumieri (Cuvier, 1830) (= *Sciaena dussumieri*) 191
Nibea soldado (Lacepède, 1802) (= *Johnius (Pseudosciaena) soldado*) 191
Scomber japonicus Houttuyn, 1782 192
Rastrelliger kanagurta (Cuvier, 1816) 193
Scomberomorus cavalla (Cuvier, 1829) 193
Thunnus thynnus (Linnaeus, 1758) 194
Thunnus albacares (Bonnaterre, 1788) 195
Thunnus alalunga (Bonnaterre, 1788) 196
Euthynnus affinis (Cantor, 1849) 196
Scomber australasicus Cuvier 1832 197
Scomber scombrus Linnaeus, 1758 197
Scomberomorus guttatus (Bloch & Schneider, 1801) 199
Scomberomorus commerson (Lacepède, 1800) 199
Acanthistius brasilianus (Cuvier, 1828) 200
Epinephelus areolatus (Forsskål, 1775) 201
Epinephelus sexfasciatus (Valenciennes, 1828) 201
Epinephelus tauvina (Forsskål, 1775) 202
Siganus rivulatus Forsskål Niebuhr, 1775 203
Siganus javus (Linnaeus, 1766) 204
Sillago sihama (Forsskål, 1775) 205
Sillago maculata Quoy & Gaimard, 1824 206
Stenotomus caprinus Jordan & Gilbert, 1882 207
Diplodus sargus (Linnaeus, 1758) 207
Sparus aurata Linnaeus, 1758 208
Sphyraena obtusata Cuvier, 1829 209
Stromateus brasiliensis Fowler, 1906 209
Pampus argenteus (Euphrasen, 1788) 210
Pampus chinensis (Euphrasen, 1788) 211
Trichiurus haumela (Forsskål, 1775) (= *Trichiurus lepturus*) 212

Contents

Lepturacanthus savala (Cuvier, 1829) 213

Xiphias gladius Linnaeus, 1758 213

Iluocoetes fimbriatus Jenyns, 1842 214

Mancopsetta maculata (Günther, 1880) 215

Cynoglossus senegalensis (Kaup, 1858) 215

Cynoglossus arel (Schneider Bloch 1801) 216

Cynoglossus lingua Hamilton, 1822 217

Paralichthys patagonicus Jordan, 1889 217

Hippoglossus hippoglossus (Linnaeus, 1758) 218

Hippoglossus slenolepis Schmidt, 1904 219

Atheresthes stomias (Jordan & Gilbert, 1880) 220

Pleuronectes platessa Linnaeus, 1758 220

Psettodes erumei (Bloch & Schneider, 1801) 222

Scaophthalmus maximus (Linnaeus, 1758) 223

Oncorhynchus kisutch (Walbaum, 1792) 224

Oncorhynchus nerka (Walbaum, 1792) 224

Oncorhynchus mykiss (Walbaum, 1792) (= *Salmo gairdneri*) 225

Micromesistius australis Norman, 1937 226

Gadus morhua Linnaeus, 1758 227

Gadus macrocephalus Tilesius, 1810 227

Melanogrammus aeglefinnus (Linnaeus, 1758) 228

Pollachius virens (Linnaeus, 1758) 229

Theragra chalcogramma (Pallas, 1814) 229

Coelorhynchus fasciatus (Günther, 1878) 230

Macruronus magellanicus Lönnberg, 1907 231

Merluccius australis (Hutton, 1872) 231

Merluccius hubbsi Marini, 1933 232

Salilota australis (Günther, 1878) 232

Notophycis marginata (Gunther, 1878) (= *Austrophycis marginata*) 233

Anguilla rostrata Lesueur, 1821 233

Bassanago albescens (Barnard, 1923) 234

Tylosurus crocodilus crocodilus (Péron & Lesueur, 1821) 234

Hyporhamphus dussumieri (Valenciennes, 1847) 235

Moolgarda seheli (Forsskål, 1775) (= *Valamugil seheli*) 236

Mugil cephalus Linnaeus, 1758 236

Plotosus lineatus (Thunberg, 1787) 237

Plotosus canius Hamilton, 1822 238

Netuma thalassina (Rüppell, 1837) 239

Abalistes stellaris (Bloch & Schneider, 1801) 239

Congiopodus peruvianus (Cuvier, 1829) 240

Pleurogammus azonus Jordan & Metz, 1913 241

Cottunculus granulosus Karrer, 1968 241

Sebastes oculatus Valenciennes, 1833 242

Prionotus nudigula Ginsburg, 1950 242

Esox lucius Linnaeus, 1758 243

Genypterus blacodes (Forster, 1801)......244
Mallotus villosus (Müller, 1776) 244
Thaleichthys pacijicus (Richardson, 1836)......245
Saurida tumbil (Bloch, 1795) 246
Harpodon nehereus (Hamilton, 1822)......246
Callorhynchus callorhynchus (Linnaeus, 1758)......247
Elasmobranchs......248
Galeocerdo cuvieri (Péron & Lesueur, 1822)......248
Scoliodon sorrokowah Thillayampallam, 1928 (= *Scoliodon laticaudus*)......248
Schroederichthys bivius (Müller & Henle, 1838) 249
Dasyatis zugei (Müller & Henle, 1841)......250
Gymnura poecilura (Shaw, 1804) 251
Dipturus chilensis Guichenot, 1848 252
Psammobatis scobina (Philippi, 1857) 252
Psammobatis normani McEachran, 1983 253
Bathyraja brachyurops (Fowler, 1910)......253
Bathyraja macloviana (Norman, 1937)......254
Bathyraja scaphiops (Norman, 1937) 255
Squalus acanthias Linnaeus, 1758......256
Discopyge tschudii Heckel, 1846......257

Chapter 10: Turtles......259
Chelonia mydas (Linnaeus, 1758)......259

Chapter 11: Mammals......261
Eumetopias jubatus (Schreber, 1776)......261
Balaena mysticetus Linnaeus, 1758......262
Erignathus barbatus (Erxleben, 1777) 262
Delphinapterus leucas (Pallas, 1776)......263
Odobenus rosmarus (Linnaeus, 1758) 264

References......267
Index 273

PREFACE

Nutritive marine fisheries resources generally account for about 16% of the protein attributed to the animal group of fish and crustaceans. This group provides high-quality sources of amino acids, which are nutritionally important types of protein that are found in only small amounts in cereals and grains. The nutrients and minerals in seafood improve brain development and reproduction. Doctors have known of strong links between fish consumption and healthy hearts because fish-eating populations have low levels of heart disease. Similarly, other groups such as phytoplankton and invertebrates possess several nutrients of importance to human health. It is always important to know the nutritional facts regarding different seafood organisms, which we have consumed from time immemorial. Although several books are available on seas and oceans, no books on the nutritional qualities of edible marine life have been available. Keeping this in consideration, an attempt has now been made. This book deals with nutritional facts of different groups of edible marine life: phytoplankton, seaweeds and marsh plants, jellyfish, crustaceans (shrimps, crabs, and lobsters), mollusks (oysters, mussels, clams, scallops, conches, abalones, limpet, squids, cuttlefish, and octopus), echinoderms (starfish, sea urchins, and sea cucumbers), prochordates (sea squirts), fish, turtles, and mammals, along with their characteristics such as classification, common names, habitats, global distribution, and biological features. The outstanding features of this publication are the easy identification of nutritionally and commercially important marine species along with their characteristic features and the nutritional facts regarding different groups of marine life. This book is of great use for undergraduate and postgraduate students belonging to the fisheries science, marine biology, aquatic science, and marine biotechnology disciplines, besides serving as a standard reference in the libraries of colleges and universities. Furthermore, because this publication provides an overview of the nutritional content of marine organisms, it is a useful reference for dietitians and doctors who are interested in knowing about the health benefits of seafood.

I am highly indebted to Dr. S. Ajmal Khan, professor (emeritus), Centre of Advanced in Marine Biology, Annamalai University, Parangipettai, India, for his valued comments and suggestions on the manuscript. I also thank Mrs. Albin Panimalar Ramesh for her help with photography.

Ramasamy Santhanam

ABOUT THE AUTHOR

Dr. Ramasamy Santhanam is the former dean of the Fisheries College and Research Institute, Tamilnadu Veterinary and Animal Sciences University, Thoothukudi, India. His fields of specialization include fisheries environment and marine biology. He is presently serving as a consultant for Advanced Aquatic Environmental Research Services (AAERS), Sultanate of Oman, and as a fisheries expert for various government and nongovernmental organizations in India. Dr. Santhanam has published 10 books on various aspects of fisheries science and 70 research papers. He was a member of the American Fisheries Society, World Aquaculture Society, and Global Fisheries Ecosystem Management Network, United States, and the International Union for Conservation of Nature's (IUCN) Commission on Ecosystem Management, Switzerland.

I

INTRODUCTION

The ocean covers approximately 70% of the Earth's surface, and it is the largest environment for living things on Earth. It is an important source of food and other resources. Although only 5% of the protein consumed by world populations comes from the marine animals of the sea, it is still an important contribution to the diet of millions of the world's inhabitants.

At present, more than 30% of humanity is suffering from malnutrition in the form of insufficient nutrient intake (undernourishment) and food-related diseases. The global magnitude and consequences of hunger and malnutrition are profound and long-lasting. Nowadays, many suffer from specific dietary micronutrient deficiencies, including deficiencies in iron, iodine, vitamin A, and zinc. In this context, marine food products derived from fish, crustaceans, mollusks, and edible aquatic plants or seaweeds could be an integral part of the human diet.

NUTRITIONAL COMPOSITION OF MARINE FOODS: FISH, CRUSTACEANS, AND MOLLUSKS

In terms of nutrient composition, marine animal food products represent one of the world's most healthy and nutritious food sources (Albert, Taconi, and Metian, 2013).

Proteins

Marine animal foods have higher protein content on an edible fresh-weight basis (mean: 17.3%) than most terrestrial meats (mean: 13.8%). Marine animal food proteins are highly digestible and have a high biological value due to their excellent essential amino acid (EAA) profile, which closely approximates the recommended human dietary EAA requirement pattern. In particular, marine animal proteins are rich dietary sources of methionine (5.9% total EAAs in mollusk proteins, 6.1% total EAAs in crustacean proteins, and 6.4% total EAAs in fish proteins) and lysine (18.2% total EAAs in mollusk proteins, 19.1% total EAAs in crustacean proteins, and 19.6% total EAAs in fish proteins). Since these EAAs are usually limited within most edible plant proteins consumed by humans, aquatic food products constitute a perfect addition to the typical plant-based diets consumed by the rural poor.

1

Fat and Fatty Acids

Marine animal foods are generally leaner on an edible fresh-weight basis (average of fat: 2.7%). Compared with terrestrial meats (average of fat: 16.6%), they have a lower saturated fat content (average of 0.16% in crustaceans, 0.32% in mollusks, and 1.19% in fish) and have a lower calorific density (average of 101.3 Kcal/100 g).

Marine animal food products contain the highest concentrations of long-chain omega-3 (n-3) polyunsaturated fatty acids of any foodstuffs, including eicosapentaenoic acid (EPA) and docosahexaenoic acid (DHA; average amounts of EPA and DHA are, respectively, 130 and 84 mg per 100 g of crustaceans, 149 and 162 mg per 100 g of mollusk,

and 279 and 467 mg per 100 g of fish). The highest levels of EPA and DHA have been reported in small pelagic fish species (average of EPA/DHA: 778/966 mg/100 g), including Atlantic mackerel, Pacific herring, Atlantic herring, and European anchovy.

In health terms, the fish-derived omega-3 (n-3) fatty acids EPA and DHA have been shown to have a positive role in infant development (including neuronal, retinal, and immune functions), cardiovascular diseases (including a reduced incidence of heart disease in adults), cancer, and various mental illnesses (including depression, attention-deficit hyperactivity disorder, and dementia).

Minerals and Trace Elements

Marine animal food products are a richer source of most essential minerals and trace elements than most terrestrial meats, including the following:

- Calcium (average of 68.2 mg/100 g in crustaceans, 39.7 mg/100 g in mollusks, and 26.0 mg/100 g in fish)
- Phosphorus (average of 230.5 mg/100 g in fish, 208.3 mg/100 g in mollusks, and 204.0 mg/100 g in crustaceans)
- Magnesium (average of 34.7 mg/100 g in crustaceans, 33.0 mg/100 g in fish, and 26.8 mg/100 g in mollusks)
- Iron (average of 3.72 mg/100 g in mollusks, 0.69 mg/100 g in fish, and 0.40 mg/100 g in crustaceans)
- Potassium (average of 367.6 mg/100 g in fish, 249.0 mg/100 g in crustaceans, and 218.8 mg/100 g in mollusks)
- Sodium (average of 394.2 mg/100 g in crustaceans, 254.9 mg/100 g in mollusks, and 73.8 mg/100 g in fish)

- Zinc (average of 11.31 mg/100 g in mollusks, 3.08 mg/100 g in crustaceans, and 0.61 mg/100 g in fish)
- Copper (average of 0.92 mg/100 g in mollusks, 0.72 mg/100 g in crustaceans, and 0.06 mg/100 g in fish)
- Manganese (average of 0.56 mg/100 g in mollusks, 0.08 mg/100 g in crustaceans, and 0.02 mg/100 g in fish)
- Selenium (average of 42.6 µg/100 g in mollusks, 41.9 µg/100 g in crustaceans, and 32.5 µg/100 g in fish)

Higher levels of mineral elements have been observed in small pelagic fish species (including European anchovy, Atlantic and Pacific herring, and Atlantic and Spanish mackerel) compared to other fish species, including calcium (average of 75 mg/100 g), iron (average of 1.8 mg/100 g), magnesium (average of 45 mg/100 g), potassium (average of 362 mg/100 g), zinc (average of 0.97 mg/100g), copper (average of 0.11 mg/100 g), manganese (0.04 mg/100 g), and selenium (mean: 38.4 µg/100 g). In addition to the mineral elements, marine animal food products are also rich dietary sources of other important essential trace elements that are generally lacking in terrestrial meat products, including iodine (<1 to 700 µg; average of 84.7 µg/100 g of 20 fish and shellfish products).

Vitamins

Marine animal food products are a richer source of several key water-soluble and fat-soluble vitamins than most terrestrial meats, including the following:

- Vitamin A (average of 263.7 IU/100 g in fish, 151.0 IU/100 g in mollusks, and 69.8 IU/100 g in crustaceans)
- Vitamin C (average of 4.0 mg/100 g in mollusks, 1.6 mg/100 g in crusta-ceans, and 0.8 mg/100 g in fish)
- Vitamin B_{12} (average of 10.0 µg/100 g in mollusks, 5.1 µg/100 g in crusta-ceans, and 3.3 µg/100 g in fish)
- Folic acid (average of 29.3 µg/100 g in crustaceans, 15.0 µg/100 g in mol-lusks, and 10.0 µg/100 g in fish)

- Vitamin E (average of 1.1 mg/100 g in fish and crustaceans and 0.80 mg/100 g in mollusks)
- Vitamin D (average of 44.9 IU/100 g in fish)
- Choline (average of 75.6 mg/100 g in crustaceans, 68.6 mg/100 g in fish, and 65.0 mg/100 g in mollusks)

Higher vitamin levels were observed in small pelagic fish species (including European anchovy, Atlantic and Pacific herring, and Atlantic mackerel) com-pared to other fish species, including riboflavin (average of 0.25 mg/100 g), niacin (7.13 mg/100 g), vitamin B_{12} (8.25 µg/100 g), vitamin A (104 IU/100 g), and vitamin D (405 IU/100 g).

SEAWEEDS

Protein

Red seaweeds such as *Porphyra* spp. (nori) have the highest levels of protein (up to 47% on a dry-weight basis); these are followed by green seaweeds, such as *Ulva lactuca* (sea lettuce), with pro-tein levels ranging between 10% and 25% on a dry-weight basis).

Amino Acids

Aspartic acid and glutamic acids con-stitute a large part of the amino acid makeup of edible seaweed proteins. Amino acids are highest within brown seaweed proteins. Moreover, edible sea-weeds such as *Palmaria palmata* (dillisk, or dulse) and *Ulva* spp. (sea lettuce) are good sources of essential amino acids such as histidine, leucine, isoleucine, methionine, and valine. The levels of isoleucine and threonine in *P. palmata* are similar to the levels found in legumes, and the histidine levels in *Ulva pertusa* are similar to the levels found in egg proteins.

3

Lipids and Fiber

The lipids present in seaweeds are rich in omega-3 polyunsaturated fatty acids, in particular EPA and to a lesser extent DHA, which are important to human health. Edible seaweeds are also a good source of dietary fiber (range: 3.4 to 9.8 g/100 g), including insoluble fiber (range: 0.5 to 2.3 g/100 g) and soluble fiber (range: 2.1 to 7.7 g/100 g on a fresh-weight basis).

Minerals and Trace Elements

Seaweeds are a rich dietary source of biologically available minerals and trace elements (compared with most other terrestrial plant food sources), including (measured as mg/100 g wet weight): iodine, 1.3–97.9 mg/100 g; iron, 3.9–45.6 mg/100 g; zinc, 0.3–1.7 mg/100 g; copper, 0.1–0.8 mg/100 g; magnesium, 78.7–573.8 mg/100 g; potassium, 62.4–2013.2 mg/100 g; and calcium, 30–575.0 mg/100 g.

Vitamins

Edible seaweeds are a rich source of many water-soluble and fat-soluble vitamins, including vitamin C (range: 8.17–184.7 mg/100 g dry weight), vitamin E (range: 0.36–17.4 mg/100 g dry weight), vitamin B_{12} (range: 1.64–78.7 g/100 g wet weight), thiamin (range: 0.14–5.04 mg/100 g dry weight), riboflavin (range: 0.14–11.70 mg/100 g dry weight), niacin (range: 0–100 mg/100 g dry weight), pyridoxine (range: 0.01–6.41 mg/100 g dry weight), inositol (range: 0.01–6.41 mg/100 g dry weight), and folic acid (range: 0–45.6 mg/100 g dry weight).

Pharmaceutical Compounds

Edible seaweeds are known to contain a variety of different species-specific bioactive chemicals with potential pharmaceutical and health-enhancing properties, including bromophenols, phytosterols, photosynthetic pigments, and immune-enhancing polysaccharides.

Among all marine food products, fish has always been considered a food necessary for good health. It is also recognized as a "brain food," owing to its importance in the development of a healthy brain. In addition, it has these benefits:

- Fish consumption reduces the risk of death from coronary heart disease, and fish consumption by women reduces the risk of suboptimal neurodevelopment in their offspring.
- Fish consumption may reduce the risk of multiple other adverse health outcomes, including ischemic stroke, nonfatal coronary heart disease events, congestive heart failure, atrial fibrillation, cognitive decline, depression, anxiety, and inflammatory diseases.
- Fish consumption may provide a greater nutritional impact than the sum of the health benefits of the individual nutrients consumed separately.

In order to alleviate the problems relating to malnutrition among the people of developing and under-developed countries, there is an urgent need to identify new species rich in nutrients from different marine biotopes. In this regard, assessment of the nutritional quality of marine flora and fauna of edible value would be of great use to add new and cheap sources of animal proteins. Further, by applying knowledge of nutritional status, one can select the needed species to harvest. This may help conserve marine ecosystems and their biodiversity, which is an urgent need presently. In the book, an attempt has been made to present detailed information on the nutritional facts of different groups of marine life. While presenting the information relating to proximate composition, mean values have been considered. Readers may refer the sources (references) concerned for detailed information.

Finally, it is important to mention that small pelagic fish species represent one of the best aquatic animal foods from a nutritional perspective.

2
PHYTOPLANKTON

Nannochloropsis oculata (Droop) D.J. Hibberd, 1981

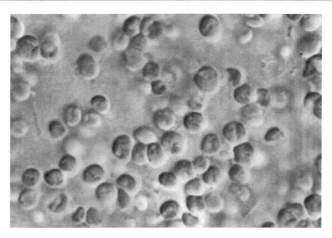

Phylum: Ochrophyta

Class: Eustigmatophyceae

Order: Eustigmatales

Family: Monodopsidaceae

Common name: Unknown

Distribution: Europe: Britain

Habitat: Marine habitats

Description: This species measures 1–2 μm in length and width. It is unicellular and free-floating. Cells are spherical and lack chlorophyll pigments other than chlorophyll *a*. This species has been proposed as a commercial source for the dietary supplement omega-3 fatty acid.

Nutritional Facts

Proximate Composition (% Dry Weight [DW])

Protein	Lipid	Carbohydrate
32.82	13.02	26.13

Source: Data from Banerjee et al. (2011).

Nannochloropsis gaditana **L.M. Lubián, 1982**

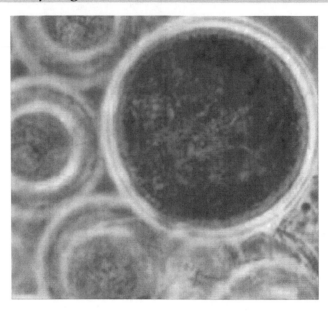

Phylum: Ochrophyta

Class: Eustigmatophyceae

Order: Eustigmatales

Family: Monodopsidaceae

Distribution: Unknown

Habitat: Marine habitats

Description: It is a small species with cells measuring about 2 to 3 µm diameter. Cells have a very simple ultrastructure with reduced structural elements. It has chlorophyll *a* and completely lacks chlorophyll *b* and *c*. It is able to build up a high concentration of a range of pigments, such as astaxanthin, zeaxanthin, and canthaxanthin.

Nutritional Facts (per 100 g Freeze-Dried Powder)

Protein	30–50%
Fat	15–30%
Minerals	10–15%
Chlorophyll *a*	1.5–3%
Total carotenoids	1.2–2%
Linoleic acid (C18:2ω3)	450–1800 mg
Linoleic acid (C18:3ω3)	17–50 mg
Eicosatrienoic acid (C20:3ω3)	800–2400 mg
EPA (C20:5ω3)	3600–10,000 mg

EPA: eicosapentaenoic acid.

Source: Data from Green Harmony Living (2014).

Phaeodactylum tricornutum Bohlin, 1897

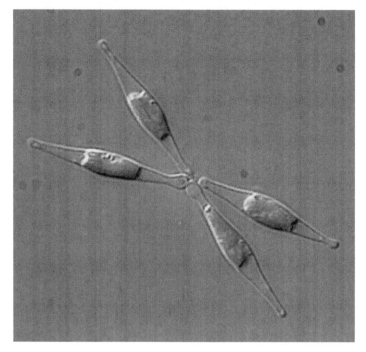

Phylum: Heterokontophyta

Class: Bacillariophyceae

Order: Naviculales

Family: Phaeodactylaceae

Common name: Unknown

Distribution: Europe and North America

Habitat: Marine habitats

Description: This species can exist in different morphotypes (fusiform, triradiate, and oval), and it can be stimulated by environmental conditions. This feature can be used to explore the molecular basis of cell shape control and morphogenesis. Unlike most diatoms, *P. tricornutum* can grow in the absence of silicon, and the biogenesis of silicified frustules is facultative.

Nutritional Facts

Proximate Composition %

Chlorophyll *a*	Protein	Cholesterol
76.7	0.41	23

Fatty Acids

(% DW)	(pg/per cell)
10.7	4.9

DW: dry weight; pg: picogram.
Source: Data from Siron et al. (1989).

9

3
SEAWEEDS AND MARSH PLANTS

GREEN ALGAE

Codium tomentosum Stackhouse, 1797

Phylum: Chlorophyta

Class: Ulvophyceae

Order: Bryopsidales

Family: Codiaceae

Common name: Velvet horn

Distribution: Northeast Atlantic Ocean; Africa

Habitat: Exposed rocks in deep rock pools on the lower shore

Description: The holdfast of this species is saucer shaped and has closely woven strands giving it a uniform appearance. The thallus or frond has a dichotomous, much-branched structure with thin branches, each with a circular cross-section. It grows to 30 cm in length and is spongy. It is covered with colorless hairs that are visible when the plant is submerged.

11

Nutritional Facts

Minerals and Proximate Composition (mg/g)

Ca	Mg	Na	K	Fat	Carbohydrate
5.5	0.8	11.8	4.3	100.2	111.5

Source: Data from Ghada and El-Sikaily (2013).

Codium iyengarii Børgesen

Phylum: Chlorophyta

Class: Ulvophyceae

Order: Bryopsidales

Family: Codiaceae

Common name: Unknown

Distribution: Southwest Asia: Pakistan

Habitat: Marine habitats

Description: This genus has thalli of two forms, either erect or prostrate. The erect plants are dichotomously branched to 40 cm long with branches forming a compact spongy structure, not a calcareous one. The final branches form a surface layer of a close palisade cortex of utricles. The nonerect species form either a prostrate or globular thallus with a velvet-like surface, with the final branches forming a close cortex of utricles.

Nutritional Facts

Proximate Composition

Protein (%)	Carbohydrate (%)	Lipid (%)
2.01	7.0	4.0

Minerals (ppm)

Calcium	Ascorbic Acid
9	Low

Source: Data from Ambreen et al. (2012).

Caulerpa lentillifera J. Agardh

Phylum: Chlorophyta

Class: Ulvophyceae

Order: Bryopsidales

Family: Caulerpaceae

Common name: Round seagrapes

Distribution: Indian and Pacific oceans

Habitat: Rocks and coral rubble

Description: This pretty seaweed resembles bunches of little grapes and is made up of tiny balls. Each "grape" is tiny (0.1–0.2 cm), usually spherical on a stalk. The "grapes" are usually tightly packed on a vertical "stem," often forming a sausage-like shape (2–10 cm long). These bunches of "grapes" emerge from a long horizontal "stem" that creeps over the surface. Color ranges from bright green to bluish and olive green.

Nutritional Facts

Proximate Composition
(g/100 g Sample Dry Basis)

Crude protein	12.49
Crude lipid	0.86
Crude fiber	3.17
Ash	24.21
Carbohydrate	59.27
Moisture	25.31

Minerals (mg/100 g Dry Basis except Cu and I in µg/100 g)

P	K	Ca	Mg	Zn	Mn	Fe	Cu	I
1030	970	780	630	2.6	7.9	9.3	2200	1424

Vitamins (mg/100 g Edible Portion)

E	C	Thiamin	Riboflavin	Niacin	Total
2.22	1.00	0.05	0.02	1.09	170

Amino Acids (g/100 g Sample Dry Basis)

Essential Amino Acids	
Threonine	0.79
Valine	0.87
Lysine	0.82
Isoleucine	0.62
Leucine	0.99
Phenylalanine	0.61
Total	4.7
Nonessential Amino Acids	
Aspartic acid	1.43
Serine	0.76
Glutamic acid	1.78
Glycine	0.85
Arginine	0.87
Histidine	0.08
Alanine	0.85
Tyrosine	0.48
Proline	0.57
Total	7.67
Total amino acids	12.37

Source: Data from Ratana-arporn and Chirapart (2006).

Caulerpa racemosa (Forsskål) J. Agardh, 1873

Phylum: Chlorophyta
Class: Ulvophyceae
Order: Bryopsidales
Family: Caulerpaceae

Common name: Sea grapes
Distribution: From Bermuda and Florida to Brazil, including all of the Caribbean.

Habitat: From shallow muddy bays to clear water reef environments, at depths from near the surface to 100 m

Description: This species has erect branches arising from a horizontal stolon attached to the sediment at intervals by descending rhizomes. The erect branches arise every few centimeters, reaching as much as 30 cm in height. A large number of branchlets, which resemble ovate or spherical bodies, arise from each erect branch. Where branches and stolons are close together, the branchlets form a dense mat of spherical structures. The plants are coenocytic (i.e., the plant is multinucleate and nonseptate).

Nutritional Facts

Proximate Composition

Moisture (% FS)	Carbohydrate (% DW)	Ash (% DW)
92.00	67.40	10.64

FS: fresh sample; DW: dry weight.

Proximate Composition (% Dry Weight)

Protein	Fiber	Lipid
10.52	11.29	0.15

Source: Data from Ahmad et al. (2012).

Amino Acids (% Dry Weight)

Aspartic acid	8.3 ± 0.2
Alanine	6.9 ± 0.5
Arginine	5.9 ± 0.7
Glutamic acid	15.3 ± 0.6
Glycine	7.2 ± 0.9
Histidine	3.3 ± 1.0
Isoleucine	3.8 ± 0.2
Leucine	7.8 ± 0.4
Lysine	7.1 ± 0.9
Methionine	1.8 ± 0.9
Phenylalanine	5.2 ± 0.2
Proline	5.8 ± 1.1
Serine	7.2 ± 1.1
Threonine	6.7 ± 1.5
Tyrosine	2.2 ± 0.4
Valine	6.9 ± 0.8

Source: Data from Rameshkumar et al. (2013).

Halimeda tuna (J. Ellis & Solander) J.V. Lamouroux, 1816

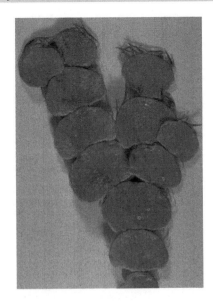

Phylum: Chlorophyta

Class: Bryopsidophyceae

Order: Bryopsidales

Family: Halimedaceae

Common name: Calcareous green seaweed, cactus algae

Distribution: Europe, North America, Central America, Caribbean, Southwest Asia, Australia, and New Zealand

Habitat: Found on hard rocky substratum less than 2 m depth

Description: Thallus is calcified and distinctly segmented with initial branching in one plane. Segments are disc-like to triangular, up to 2 cm wide.

Internodal siphons are uncalcified, are united in twos or threes, and terminate in pseudodichotomous laterals. Surface cells are appressed to one another in a honeycomb pattern, 25–75 μm in diameter.

Nutritional Facts

Proximate Composition

Protein (%)	Carbohydrate (%)	Lipid (%)
2.0	10.0	4.0

Minerals (ppm)

Calcium	Ascorbic Acid
15	33

Source: Data from Ambreen et al. (2012).

Rhizoclonium implexum (Dillwyn) Kützing, 1845

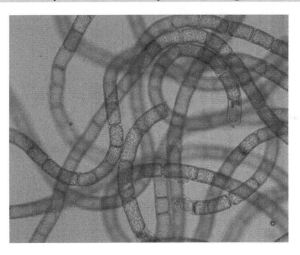

Phylum: Chlorophyta

Class: Ulvophyceae

Order: Cladophorales

Family: Cladophoraceae

Common name: Green tidal-flat mats

Distribution: Cosmopolitan (Southern Australia, from Venus Bay, South Australia, to Victoria)

Habitat: Intertidal or sheltered waters

Description: Thalli are in small floating mats, entangled with other algae such as *Enteromorpha*. Plants consist of thin mats of loose, fine, entangled threads. Filaments are unbranched, and very rarely it has unicellular rhizoidal branches. Cells are cylindrical.

Nutritional Facts

Proximate Composition

Protein (%)	Carbohydrate (%)	Lipid (%)
5.6	14.4	8.0

Minerals (ppm)

Calcium	Ascorbic Acid
14	33

Source: Data from Ambreen et al. (2012).

Cladophora glomerata (L.) Kutzin 1843

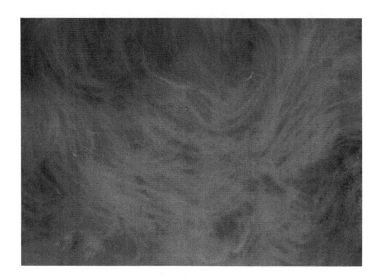

Phylum: Chlorophyta

Class: Ulvophyceae

Order: Cladophorales

Family: Cladophoraceae

Common name: Not available (NA)

Distribution: Baltic Sea, Gulf of Finland, Gulf of Mexico, Gulf of Riga, Mediterranean Sea, and North Atlantic Ocean

Habitat: Attached to rocks or timbers submerged in shallow lakes and streams

Description: Coarse in appearance, with regular-branching filaments that have cross walls separating multinucleate segments. It grows in the form of a tuft or ball with filaments that may range up to 13 cm in length. Asexual reproduction involves small, motile spores (zoospores) with four flagella; and in sexual reproduction, the biflagellate gametes normally unite. They may sometimes develop into new plants without union.

Nutritional Facts

Proximate Composition

Crude Ash (%)	Crude Oil (%)	Protein (%)
2.44	2.48	14.13

Source: Data from Akköz et al. (2011).

Ulva clathrata (Roth) C. Agardh 1811 = *Enteromorpha clathrata*

Phylum: Chlorophyta

Class: Ulvophyceae

Order: Ulvales

Family: Ulvaceae

Common name: Bright-green nori

Distribution: Worldwide in temperate moderate to calm seas

Habitat: Intertidal and shallow waters

Description: This species is light green in color and is 2–8 cm in height. The thin cylindrical threads are 1–3 mm in width and are cylindrical or slightly flat.

Nutritional Facts

Proximate Composition

Crude protein	20–26%
Essential amino acids	32–36% (of crude protein)
Saturated fatty acids	31%

Minerals

Ca (g/kg)	Fe (g/kg)	Cu (mg/kg)	Zn (mg/kg)	As (mg/kg)
9	0.6	44	93	2

Source: Data from Rodriguez et al. (2011).

Ulva pertusa Kjellman, 1897

Phylum: Chlorophyta

Class: Ulvophyceae

Order: Ulvales

Family: Ulvaceae

Common name: Sea lettuce

Distribution: Indo-Pacific Ocean

Habitat: Lower littoral and upper subtidal zones of a wide variety of habitats: on rocks, in pools, on other marine organisms

Description: Thallus forms a distromatic blade. It grows up to 20 cm long and is irregularly orbicular, lobed, thick, and tough when small, but oval and more delicate toward the margins when large. Blade is bright to dark green, glossy, without marginal teeth, and generally with perforations of variable size and irregular shape. Basal part of the blade is cuneate and thick (up to 500 µm). It is without central cavity but with characteristic concentric wrinkles around the holdfast.

Ulva intestinalis Linnaeus

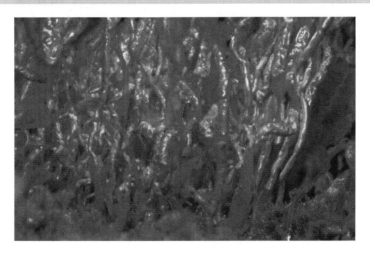

Phylum: Chlorophyta

Class: Ulvophyceae

Order: Ulvales

Family: Ulvaceae

Common name: Gutweed, grass kelp

Distribution: Worldwide, except for in polar waters

Habitat: Wide range of habitats on all levels of the shore

Description: It has inflated, irregularly constricted, tubular, generally unbranched fronds that grow out from a small base. The fronds may be 100–300 mm in length and have a diameter of 6–18 mm. They have rounded tips as well. The plant is a summer annual and decays to become a bleached and decaying mass toward the end of the summer season.

Nutritional Facts

Proximate Composition

	Protein	Lipid	Ash	Moisture
U. pertusa	15.4	4.8	27.2	6.0
U. intestinalis	17.9	8.0	27.6	6.3

Dietary Fiber (% Dry Weight)

U. pertusa	27.8
U. intestinalis	32.4

Macro Elements (mg/100 g Dry Weight)

	U. pertusa	U. intestinalis
Mg	3,670.0	3,098.1
K	1,224.1	2,538.6
Ca	669.4	1,047.8
Na	376.7	1,064.5
C	146.1	1,705.5
P	177.0	271.9

Trace Elements (mg/100 g Dry Weight)

	U. pertusa	U. intestinalis
Cu	1.0	0.9
Zn	0.8	1.5

Amino Acids (mg/100 mg Dry Weight)

	U. pertusa	U. intestinalis
Aspartic acid	1.19	1.46
Serine	0.58	0.65
Glutamic acid	1.17	1.42

Amino Acids (mg/100 mg Dry Weight)

	U. pertusa	U. intestinalis
Glycine	0.66	0.74
Histidine	0.13	0.13
Arginine	0.62	0.57
Threonine	0.54	0.72
Alanine	0.92	1.19
Proline	0.54	0.50
Tyrosine	0.35	0.28
Valine	0.60	0.70
Lysine	0.46	0.37
Isoleucine	0.40	0.43
Leucine	0.80	0.85
Phenylalanine	0.56	0.62

Essential Amino Acids (mg/g Protein)

	U. pertusa	U. intestinalis
Leucine	52	49.7
Isoleucine	25.9	25.3
Lysine	30.1	19.5
Threonine	34.8	41.7
Tyrosine + phenylalanine	59.5	52.0

Source: Data from Benjama and Masniyom (2011).

Ulva faciata Delile 1813

Phylum: Chlorophyta
Class: Ulvophyceae
Order: Ulvales

Family: Ulvaceae
Common name: Sea lettuce

Distribution: Worldwide: Eastern Atlantic Ocean, Caribbean, and Indian and Pacific Oceans

Habitat: On intertidal rocks, in tide-pools, and on reef flats

Description: Thalli, which are up to 1 m long, are thin and sheet-like, consisting of wide blades. They are 10–15 cm wide at the base, tapering upward to less than 2.5 cm wide at the tip. Basally they are broadened, but upper portions are divided deeply into many ribbon-like segments. Margins are smooth and often undulate. Holdfast is small without dark rhizoids. Color of the thalli is bright grass green to dark green and gold at the margins when reproductive.

Nutritional Facts

Proximate Composition (% Dry Weight)

Protein	Lipids	Carbohydrate
14.7	0.5	70.1

Amino Acids (% Dry Weight)

Aspartic acid	10.7
Alanine	9.2
Arginine	6.3
Glutamic acid	11.9
Glycine	6.2
Histidine	3
Isoleucine	3.2
Leucine	8.6
Lysine	5.9
Methionine	1.3
Phenylalanine	5.5
Proline	3.9
Serine	6.7
Threonine	4.9
Tyrosine	4.7
Valine	6.3

Source: Data from Rameshkumar et al. (2013).

Ulva lactuca Linnaeus

Phylum: Chlorophyta

Class: Ulvophyceae

Order: Ulvales

Family: Ulvaceae

Common name: Sea lettuce

Distribution: Worldwide: Europe, North America, Central America, Caribbean Islands, South America, Africa, Indian Ocean islands, Southwest Asia, China,

Pacific Islands, Australia, and New Zealand

Habitat: Rock and in lower-shore rock pools, and in the shallow subtidal zones

Description: Thallus is sheet-like, light green, rather delicate, and translucent. It is 250 mm long. Its color may range from light yellowish-green to darker green, but it is most commonly a vivid green underwater. The soft frond grows as a single, irregular, but somewhat round-shaped blade with slightly ruffled edges that are often torn. There can be numerous small holes or perforations scattered throughout. The frond is connected to rocks with a small, almost invisible discoid holdfast.

Nutritional Facts

Proximate Composition (g/100 g Dry Weight [DW])

Lipid	Protein	Fiber	DW	Ash
0.99	10.69	5.60	5.96	18.03

Amino Acids (mg/g Protein)

Aspartic acid	49.7
Glutamic acid	70.7
Serine	28.7
Glycine	39.7
Threonine	31.1
Arginine	37.9
Alanine	43.3
Tyrosine	23.4
Histidine	15.2
Valine	39.2
Methionine	5.9
Isoleucine	21.7
Phenylalanine	28.4
Leucine	45.1
Lysine	25.4
Proline	37.9
Essential amino acids	258.1
Nonessential amino acids	285.2
Total	543.3

Fatty Acids (% of Total Fatty Acids)

Saturated	48.34
Monounsaturated	5.11
Polyunsaturated	24.84

Source: Data from Tabarsa et al. (2012).

Ulva reticulata Forsskål, 1775

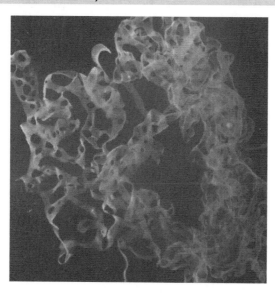

Phylum: Chlorophyta

Class: Ulvophyceae

Order: Ulvales

Family: Ulvaceae

Common name: Ribbon sea lettuce

Distribution: Indo-west Pacific region

Habitat: Rocky substrates

Description: Mature thalli of this species have irregular shapes. They are light to dark green in color, forming masses of perforated blades from a few centimeters in size to about a meter across. Juveniles are usually attached with a small discoid holdfast, becoming detached into free-living individuals.

Nutritional Facts

Proximate Composition (g/100 g Sample Dry Basis)

Protein	Lipid	Fiber	Ash	Carbohydrate	Moisture
21.06	0.75	4.84	17.58	55.77	22.51

Minerals (mg/100 g Dry Basis except Cu and I in µg/100 g)

P	K	Ca	Mg	Zn	Mn	Fe	Cu	I
180	1540	140	140	3.3	48.1	174.8	600	1124

Vitamins (mg/100 g Edible Portion)

E	C	Thiamin	Riboflavin	Niacin	Total
0	0	0.01	0.13	0	167

Amino Acids (g/100 g Sample Dry Basis)

Essential Amino Acids	
Threonine	1.15
Valine	1.34
Lysine	1.28
Isoleucine	0.90
Leucine	1.68
Phenylalanine	1.12
Total	7.47
Nonessential Amino Acids	
Aspartic acid	2.66
Serine	1.36
Glutamic acid	2.76
Glycine	1.38
Arginine	1.84
Histidine	0.23
Alanine	1.72
Tyrosine	0.77
Proline	1.08
Total	13.8
Total amino acids	21.27

Source: Data from Shanmugam and Palpandi (2008).

BROWN ALGAE

Ascophyllum nodosum (L.) Le Jolis

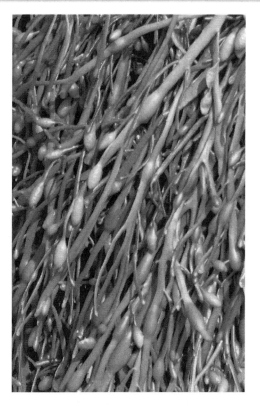

Phylum: Ochrophyta

Class: Phaeophyceae

Order: Fucales

Family: Fucaceae

Common name: Knotted wrack

Distribution: North Atlantic basin

Habitat: Coastal habitats from sheltered estuaries to moderately exposed coasts; dominant in the intertidal zone

Description: It is a brown seaweed forming single bladders centrally in long, strap-like fronds. The fronds hang downward. A number of fronds grow from each basal holdfast, and it normally regenerates new fronds from the base when one of the larger fronds is damaged.

Nutritional Facts

Proximate Composition

Protein (% Dry Mass)
3 ± 15

Fiber (g/100 g Wet Weight)

Total Fiber	Soluble Fiber	Insoluble Fiber	Carbohydrate
8.8	7.5	1.3	13.1

23

Minerals (mg/100 g Wet Weight)

Ca	K	Mg	Na	Cu	Fe	I	Zn
575	765	225	1174	0.8	14.9	18	NA

Vitamins (mg per 8 g Dry Portion)

B_1	B_2	B_3	B_6	B_8	B_9	C	E	B_{12}
0.22	0.06	0	0.001	0.001	3.65	0.65	0.03	0.13

Source: Data from MacArtain et al. (2007).

Himanthalia elongata (Linnaeus) S.F. Gray

Phylum: Ochrophyta

Class: Phaeophyceae

Order: Fucales

Family: Himanthaliaceae

Common name: Buttonweed, thongweed

Distribution: Baltic Sea, North Sea and northeast Atlantic Ocean

Habitat: Gently shelving rocky shores in the lower littoral zone and the sublittoral zone, particularly on shores with moderate wave exposure

Description: Thallus consists of a button-shaped vegetative thallus that is 30 mm wide and 25 mm high. It is long, narrow, strap-like, and sparingly branched. The light yellow-brown reproductive receptacle is 2 m in length and up to 10 mm in width, on which the conceptacles are borne.

Nutritional Facts

Fiber (g/100 g Wet Weight)

Total Fiber	Soluble Fiber	Insoluble Fiber	Carbohydrate
9.8	7.7	2.1	15.0

Fiber and Carbohydrate (mg per 8 g)

Total Fiber	Soluble Fiber	Insoluble Fiber	Carbohydrate
2.6	2.1	0.6	4.0

Minerals (mg/100 g Wet Weight)

Ca	K	Mg	Na	Cu	Fe	I	Zn
30	1351	90	601	0.1	5.0	11	1.7

Polyunsaturated Fatty Acids (PUFA)
(% of Total Fatty Acid)

Saturated	Monounsaturated	PUFA
39.06	22.75	38.16

Source: Data from MacArtain et al. (2007).

Hormophysa cuneiformis (Gmelin) Silva

Phylum: Ochrophyta

Class: Phaeophyceae

Order: Fucales

Family: Sargassaceae

Common name: Hallow-green nori

Distribution: Indian Ocean, Kenya, Madagascar, Mozambique

Habitat: Rocks in lower intertidal zones along shorelines with a frequent gentle current

Description: The plants have a filamentous holdfast and triquetrous main branches. The vesicles are oblong to elliptical. The wing-like leaves are with serrations on the margin. The color changes to a dark brown when dried. The plant grows on rocks in lower intertidal zones along shorelines with a frequent gentle current.

Nutritional Facts

Proximate Composition

Moisture (% FS)	Carbohydrate (% DW)	Ash (% DW)
86.86	40.57	26.81

Proximate Composition (% Dry Weight)

Protein	Fiber	Lipid
6.42	25.36	0.84

FS: fresh sample; DW: dry weight.
Source: Data from Ahmad et al. (2012).

25

Sargassum tenerrimum J. Agardh 1848

Phylum: Ochrophyta

Class: Phaeophyceae

Order: Fucales

Family: Sargassaceae

Common name: NA

Distribution: Throughout the temperate and tropical oceans of the world

Habitat: On the rocks in the tidal ditch or tidal pool in the intertidal zone

Description: Plants, which reach a height of 30–40 cm, are delicate and pyramidal in form. They have a disc-shaped holdfast and are yellowish-brown in color. Axis is glabrous and rounded. Ultimate branchlets are modified into vesicles and receptacles. Leaves, which are alternately arranged, are thin, transparent, and 2–6 cm long and 0.5–1.5 cm broad. Vesicles are stalked and spherical. Receptacles are richly branched and spinose.

Nutritional Facts

Proximate Composition (%)

Protein	Carbohydrate	Lipid
12.42	23.55	1.46

Minerals (ppm)

Ca	Cu	Fe	K	Mg	Na	Zn
11703	47	623	105217	9732	35933	14518

Source: Data from Rizvi and Shameel (2010).

Sargassum variegatum

Phylum: Ochrophyta

Class: Phaeophyceae

Order: Fucales

Family: Sargassaceae

Common name: NA

Distribution: Tropical areas of the world

Habitat: Near-shore areas

Description: This species may grow to a length of several meters. It is generally brown or dark green in color and consists of a holdfast, a stipe, and a frond. Oogonia and antheridia occur in conceptacles embedded in receptacles on special branches. It has a rough sticky texture, which together with a robust but flexible body helps it withstand strong water currents.

Nutritional Facts

Proximate Composition

Protein (%)	Carbohydrate (%)	Lipid (%)
3.0	36.0	1.0

Minerals (ppm)

Calcium	Ascorbic Acid
8	51

Source: Data from Ambreen et al., 2012

Sargassum wightii Greville

Phylum: Ochrophyta

Class: Phaeophyceae

Order: Fucales

Family: Sargassaceae

Common name: NA

Distribution: Tropical areas of the world

Habitat: Near-shore areas

Description: This species has a disc-shaped holdfast and its main branch, which is flat and 3 mm wide, is without spines. Secondary lateral branches alternate at intervals of 2–3 cm. Leaves are thick and lanceolate to linear-lanceolate.

27

They are 4 cm long and 6 mm wide. Androgynous receptacles are terete or compressed, about 5 mm long, with small spines on the apical part.

Nutritional Facts

Proximate Composition

Protein	Lipid	Carbohydrate	Alginic Acid
3.15–7.20	0.15–1.55	6.65–15.18	12.10–26.32

Source: Data from Jayasankar (1993).

Sargassum polycystum C. Agardh

Phylum: Ochrophyta

Class: Phaeophyceae

Order: Fucales

Family: Sargassaceae

Common name: Rough-stemmed sargassum, sargassum weed

Distribution: Indian Ocean, North Atlantic Ocean

Habitat: On rocks in lower intertidal zones in relatively calm waters

Description: The erect branches of this species have numerous spines on the stem. The creeping branches found post maturation are smooth on the edges and form secondary branches with secondary holdfasts on the terminal portions. Leaves are lanceolate to oblong with serrations, and vesicles are spherical. The plants are 1–2 m high, and they form large communities on rocks in lower intertidal zones in relatively calm waters.

Nutritional Facts

Proximate Composition

Moisture (% FS)	Carbohydrate (% DW)	Ash (% DW)
83.51	34.93	21.87

FS: fresh sample; DW: dry weight.

28

Proximate Composition (% Dry Weight)

Protein	Fiber	Lipid
7.78	34.71	0.71

Source: Data from Ahmad et al. (2012).

Turbinaria conoides (J. Agardh) Kutzing 1860

Phylum: Ochrophyta

Class: Phaeophyceae

Order: Fucales

Family: Sargassaceae

Common name: Agar-agar Lesong

Distribution: Indian Ocean

Habitat: Intertidal zone

Description: The thallus of this species is erect and 13 cm tall. It is dark brown in color and is attached to substrate with a branched holdfast. The erect axis is subcylindrical, supporting alternatively and polystichously arranged stalked turbinate leaves. The thin triangular leaves are 13–20 mm long, are concave at the center, and have single margins with large sharp serrations.

Nutritional Facts

Proximate Composition

Moisture (% FS)	Carbohydrate (% DW)	Ash (% DW)
83.79	41.03	21.37

FS: fresh sample; DW: dry weight.

Proximate Composition (% DW)

Protein	Fiber	Lipid
7.40	29.61	0.59

DW: dry weight.

Amino Acids (μg/g Dry Weight)

Aspartic acid	12.00
Glutamic acid	17.50
Asparagine	5.733
Serine	15.53
Glutamine	0.753
Glycine	5.133
Threonine	6.667
Arginine	11.37
Alanine	3.767
Cystine	3.367
Tyrosine	2.267
Histidine	10.97
Valine	4.100
Methionine	20.93
Isoleucine	17.70

Phenylalanine	5.567
Leucine	20.67
Lysine	7.667
Proline	3.33
Tryptophan	0.012
Taurine	0.010

Fatty Acids (μg/g Dry Weight)

Palmitic acid	21.55
Stearic acid	16.67
Oleic acid	14.47
Linolenic acid	20.31
Alpha linolenic acid	11.60
Moroctic acid	6.093

Source: Data from Sridharan and Dhamotharan (2012).

Calpomenia sinuosa (Mertens ex Roth) Derbès & Solier, 1851

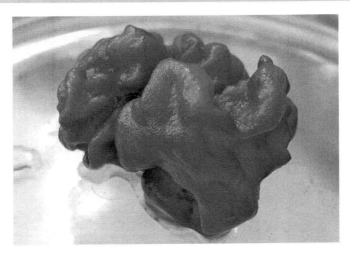

Phylum: Ochrophyta

Class: Phaeophyceae

Order: Ectocarpales

Family: Scytosiphonaceae

Common name: Sea potato, sea balloon

Distribution: Florida, Gulf of Mexico, the Caribbean

Habitat: Lower intertidal to 15 m deep; firmly attached to hard surfaces or epiphytic on other organisms

Description: Plants are smooth, hemispherical, irregularly lobed, golden brown, and hollow. This species grows to 30 cm diameter and 10 cm high, with multiple attachments to substrate. It is often covered with fine colorless hairs. Reproductive sori are seen as dark raised patches on the surface. Thallus is membranaceous and 300–500 μm thick with 4–6 cell layers.

Nutritional Facts

Minerals and Proximate Composition (mg/g)

Ca	Mg	Na	K	Fat	Carbohydrate
3.8	1.1	24.5	9.2	144.7	118.0

Source: Data from Ghada and El-Sikaily (2013).

Chnoospora minima (Hering) Papenfuss, 1956

Phylum: Ochrophyta

Class: Phaeophyceae

Order: Ectocarpales

Family: Scytosiphonaceae

Common name: Hornwort, coontail

Distribution: Tropical and subtropical Indian, Pacific, and western Atlantic oceans

Habitat: Subtidal zones

Description: Fronds are erect from a discrete base. Silky colorless hairs are common along the middle and upper parts of plants. This species grows high in the intertidal region, usually on vertical or sloping igneous rocks.

Nutritional Facts

Proximate Composition (% Dry Weight)

Protein	Lipids	Carbohydrate
11.3	0.9	28.5

31

Amino Acids (% Dry Weight)

Aspartic acid	10.9	Lysine	6.4
Alanine	8.5	Methionine	3.1
Arginine	4.0	Phenylalanine	4.7
Glutamic acid	13.2	Proline	4.3
Glycine	5.6	Serine	6.7
Histidine	2.6	Threonine	5.4
Isoleucine	4.6	Tyrosine	2.5
Leucine	7.3	Valine	5.3

Source: Data from Rameshkumar et al. (2013).

Dictyota dichotoma var. *intricata* (C. Agardh) Greville, 1830

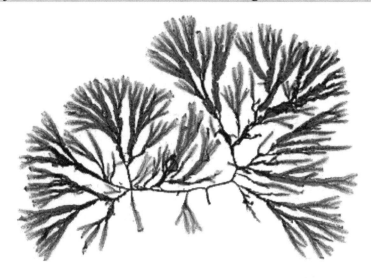

Phylum: Ochrophyta

Class: Phaeophyceae

Order: Dictyotales

Family: Dictyotaceae

Common name: Common forked tongue

Distribution: Worldwide

Habitat: Shallow water to 21 m, on hard surfaces

Description: This species possesses medium-brown flat blades that branch dichotomously. Plants are often iridescent blue when they are underwater. Sporangia are in scattered patches on blades. This species has relatively long distances between branching. The lower branches may be occasionally fringed if the plant has been heavily grazed.

Nutritional Facts

Proximate Composition

Protein (%)	Carbohydrate (%)	Lipid (%)
5.0	2.8	6.8

Mineral and Vitamin (ppm)

Calcium	Ascorbic Acid
10	52

Source: Data from Ambreen et al. (2012)

Dictyota indica Sonder ex Kützing

Phylum: Ochrophyta

Class: Phaeophyceae

Order: Dictyotales

Family: Dictyotaceae

Common name: Strap brown seaweed

Distribution: North America, Caribbean Islands, South America, Africa, Southwest Asia, Southeast Asia, Australia, and New Zealand

Habitat: Reef environments; on rocky substrates

Description: This species has branches that fork near their ends. The tips may be rounded or pointed. Generally, it forms mats of dense to loose-packed flat leaves that overgrow the substrate. Color of the species is light to medium brown and/or green to blue-green, occasionally with bright blue tints.

Nutritional Facts

Proximate Composition

Protein (%)	Carbohydrate (%)	Lipid (%)
7.41	11.0	2.0

Mineral and Vitamin (ppm)

Calcium	Ascorbic Acid
80	32

Source: Data from Ambreen et al. (2012).

Padina australis Hauck

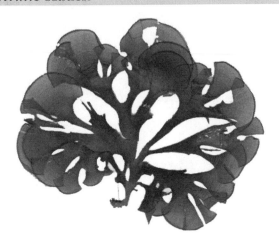

Phylum: Ochrophyta

Class: Phaeophyceae

Order: Dictyotales

Family: Dictyotaceae

Common name: Fan-leaf seaweed

Distribution: Japan, Korea, China, Hong Kong, Philippines, Vietnam, Singapore, Malaysia, Papua New Guinea, Fiji, Solomon Islands, Hawaii, Taiwan, Thailand, Bangladesh, India, Indonesia, and Kuwait

Habitat: Tidal pool or 1–2 m depth of water

Description: Thallus is yellow-brown in color in fresh and dried plants. It is upright with a rhizoid holdfast. It is fan shaped with a membrane-like texture. It measures 6–18 cm in height and 5–15 cm in width. Proliferations are seen on the hair bands at the older part of the thallus. White hairs are in concentric arrangement on two surfaces of the thallus, and brown fluffs are present over the base to the middle part of the thallus.

Padina durvillei Bory Saint-Vincent

Phylum: Ochrophyta

Class: Phaeophyceae

Order: Dictyotales

Family: Dictyotaceae

Common name: NA

Distribution: Tropical and temperate seas

Habitat: Lower rock pools and reef edges

Description: This species has fan-shaped fronds, which are entire when young but dissected when older. Distinctive, flattened fan-shaped thallus is seen with concentric markings and rolled edges. It attaches to rock by rhizoids. It is branched only in one plane, with thin fronds, often lacerated from edge to base.

Padina fernandeziana Skottsberg & Levring, 1941

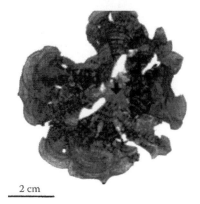

2 cm

Phylum: Ochrophyta

Class: Phaeophyceae

Order: Dictyotales

Family: Dictyotaceae

Common name: NA

Distribution: Southeast Pacific: endemic to Juan Fernández Islands

Habitat: Benthic; hard substrates, rocks in shallow water

Description: It has leafy, fan-like blades. The leaves are lightly calcified and are about 5 cm in diameter. This species lays down the calcium carbonate on the cell wall surface rather than between the cells (intercellularly).

Padina gymnospora (Kutzing) Sonder

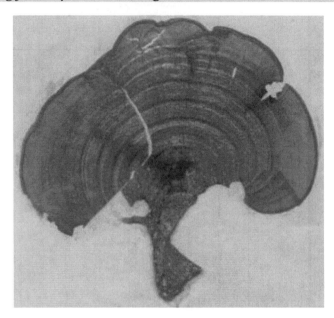

Phylum: Ochrophyta

Class: Phaeophyceae

Order: Dictyotales

Family: Dictyotaceae

Common name: Limey petticoat

Distribution: Japan, Korea, China, Hong Kong, Philippines, Taiwan, and India

Habitat: In 2–3 m depth of water

Description: Thallus is yellow-brown in color and becomes dark brown when dried. It is upright with a rhizoid holdfast and is fan shaped with a thick membrane-like texture. Thallus measures 5–20 cm in height and 4–17 cm in width. Concentric hair bands are 1–2 mm in width. Thallus is 85–150 µm in thickness at the margin.

Nutritional Facts

Proximate Composition (% Dry Weight)

Species	Ashes	Protein	Carbohydrate	Lipids	Fibers
P. australis	5.50	1.50	—	0.80	9.60
P. durvillei	34.43	5.24	44.18	0.69	7.57
P. fernandeziana	35.75	8.31	44.07	2.01	44.61
P. gymnospora	36.61	9.86	1.86	0.11	9.07

Source: Data from Goecke et al. (2012).

Amino Acids (% Dry Weight)

Aspartic acid	12.7 ± 1.3
Alanine	6.7 ± 0.3
Arginine	4.7 ± 0.2
Glutamic acid	13.9 ± 1.2
Glycine	7.0 ± 0.7
Histidine	2.7 ± 0.5
Isoleucine	5.3 ± 0.6
Leucine	7.9 ± 0.4

Lysine	6.2 ± 0.4
Methionine	1.5 ± 0.5
Phenylalanine	5.3 ± 0.3
Proline	4.0 ± 0.5
Serine	6.7 ± 0.4
Threonine	5.7 ± 0.3
Tyrosine	2.9 ± 0.6
Valine	6.5 ± 0.6

Source: Data from Rameshkumar et al. (2013).

Padina pavonica (Linnaeus) J.V. Lamouroux

Phylum: Ochrophyta

Class: Phaeophyceae

Order: Dictyotales

Family: Dictyotaceae

Common name: Peacock's tail

Distribution: Southern coast of England, southward to the Mediterranean

Habitat: Lower intertidal in pools; soft rock, sand, and detritus

Description: Thallus is fan shaped and becomes funneled when older. It measures 120 mm long and 10–100 mm broad.

Fronds are thin and translucent. Color is olive to yellow-brown and darker near the base. Apices are ringed with brownish or silvery hairs.

Nutritional Facts

Proximate Composition and Polyunsaturated Fatty Acids (PUFA) (Wet Weight %)

Protein	Lipid	Moisture	Ash	PUFA
1.65	0.87	71.6	3.7	26.2

Source: Data from Polat and Ozogul (2013).

37

Padina vickersiae Hoyt in Britton & Millspaugh 1920

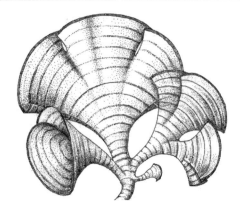

Phylum: Ochrophyta

Class: Phaeophyceae

Order: Dictyotales

Family: Dictyotaceae

Common name: NA

Distribution: Atlantic Islands, North America, Central America, Caribbean Islands, South America, Africa, and Southwest Asia

Habitat: Shallow waters; associated with mangrove roots and submerged rocks, shells, or coral fragments

Description: This species is thin and somewhat ruffled. It has a fan-shaped appearance and light brown color. Whole plants form tufts up to 15 cm in height with fan-like blades, segmented at the margins and arising from short, fibrous stalks. Holdfasts are bulbous and rhizoidal.

Nutritional Facts

Proximate Composition (% Dry Weight)

Ash	Protein	Lipid	Fiber
21.74	18.62	1.43	9.78

Source: Data from Goecke et al. (2012).

Padina tetrastromatica Hauck

Phylum: Ochrophyta

Class: Phaeophyceae

Order: Dictyotales

Family: Dictyotaceae

Common name: NA

Distribution: Tropical and subtropical regions

Habitat: Mangrove swamps (attached to mud), or intertidal

Description: Thalli are flabelliform and are usually divided into several small lobes. This species is easily recognized due to dark double lines of sporangia. Blades are composed of two layers of cells.

Nutritional Facts

Amino Acids (µg/g Dry Weight)

Aspartic acid	27.083
Glutamic acid	42.212
Asparagine	59.200
Serine	26.000
Glutamine	14.698
Glycine	33.013
Threonine	17.931
Arginine	24.562
Alanine	25.150
Cystine	26.995
Tyrosine	18.671
Histidine	44.898

Valine	45.332
Methionine	29.826
Isoleucine	74.393
Phenylalanine	50.595
Leucine	49.535
Lysine	5.022

Vitamins (µg/g Dry Weight)

B₆	B₂	B₁	Nicotinic Acid
115	75	46	68

Fatty Acids (µg/g Dry Weight)

Caprylic acid	22.0
Lauric acid	67.0
Tridecanoic acid	27.0
Myristic acid	11.067
Pentadecanoic acid	11.6
Palmitaloic acid	34.1
Heptadecanoic acid	27.3
Stearic acid	29.4
Oleic acid	24.0
Linoleic acid	47.7
α Linoleic acid	10.0
γ Linoleic acid	8.6
Palmitic acid	5001

Minerals (µg/g Dry Weight)

Cu	Zn	Mn	Mg	Na	K	Ca
8.641	19.92	8.75	9.6	39.11	29.65	2053.4

Source: Data from Goecke et al. (2012).

Stypopodium schimperi (Kützing) M. Verlaque & Boudouresque 1991

Phylum: Ochrophyta

Class: Phaeophyceae

Order: Dictyotales

Family: Dictyotaceae

Common name: Upside-down jellyfish

Distribution: Eastern Mediterranean and the Red Sea

Habitat: Well-lit areas on rocky substrates in depths of 0–20 m

Description: This species has thin, fan-shaped appendages that have longitudinal tears. Their appearance is almost transparent with a brown coloration. The appendages are covered in rows of hair, and they are without any calcification. The reproductive spores are located between the hairs. This species reaches a height of 100–300 mm.

Nutritional Facts

Proximate Composition (%)

Protein	Lipid	Moisture	Ash	Polyunsaturated Fatty Acids
2.37	2.16	78.4	3.8	15.9

Source: Data from Polat and Ozogul (2013).

Eisenia arborea Aresch 1876

Phylum: Ochrophyta

Class: Phaeophyceae

Order: Laminariales

Family: Lessoniaceae

Common name: Southern sea palm

Distribution: Eastern Pacific from Vancouver Island, Canada, south to Isla Magdalena, Mexico

Habitat: Intertidal zones

Description: This is an edible seaweed, a source of nutrients for grazing marine invertebrates, and a source of alginic acid, a food thickener. This species has a large, thick stipe above its holdfast with two branches terminating in multiple blades.

Nutritional Facts

Proximate Composition (%, dw)

Moisture	Protein	Ash	Lipids	Carbohydrate	Energy(kJ/g)
10.3	9.4	24.8	0.6	49	9.8

Source: Data from Pennywhite (2007).

Laminaria digitata (Hudson) J.V. Lamouroux

Phylum: Ochrophyta

Class: Phaeophyceae

Order: Laminariales

Family: Laminariaceae

Common name: Oarweed

Distribution: Coasts of Britain and Ireland

Habitat: Lower intertidal and shallow subtidal areas; grows on rock

Description: This species is dark brown and is 2 m in length. It has a claw-like holdfast, and a smooth, flexible stipe that is oval in cross-section and does not snap easily when bent. A laminate blade that is 1.5 m long splits into finger-like segments.

Nutritional Facts

Proximate Composition

Crude protein (g/100 g dry weight)	6.2 ± 0.4

Amino Acids (mg/g Dry Weight)

Aspartic acid	4.69
Glutamic acid	3.86
Serine	2.45
Threonine*	3.41
Glycine	3.31
Alanine	4.51
Arginine	2.96
Proline	1.91
Valine*	6.01
Methionine*	1.49
Isoleucine*	2.61
Leucine*	4.45
Tryptophan*	0.19
Phenylalanine*	2.82
Cystine	1.96
Lysine*	4.77
Histidine*	2.38
Tyrosine	1.74
Total	55.46

Essential amino acids.

41

Vitamins (mg/100 g Dry Weight)

B_1	0.24
B_2	0.85
B_6	0.09
Niacin	1.58

Minerals (mg/100 g Dry Weight)

Ca	880
Mg	550
P	300
I	170
Na	2532

K	5951
Ni	0.325
Cr	0.227
Se	<0.05
Fe	1.19
Zn	0.886
Mn	0.294
Cu	0.247
Pb	0.087
Cd	0.017
Hg	0.054
As	0.087

Source: Data from Kolb et al. (2004).

Saccharina latissima (Linnaeus) J.V. Lamouroux (= *Laminaria saccharina*)

Phylum: Ochrophyta

Class: Phaeophyceae

Order: Laminariales

Family: Laminariaceae

Common name: Sugar kelp

Distribution: Circumboreal, northern Russia south to Galicia, Spain; Britain and Ireland

Habitat: Intertidal pools and shallow subtidal regions

Description: It is yellow-brown and it 3 m in length. A claw-like holdfast; a small, smooth, flexible stipe; and an undivided laminate blade are present. The frond is characteristically dimpled with regular bullations (depressions).

Nutritional Facts

Proximate Composition (% Dry Weight)

Protein	Lipid	Carbohydrate	Minerals
6–15	1–3	50–65	15–40

Organic Residue (% Dry Weight)

Alginic Acid	Laminarine	Mannitol	Cellulosic Content
18	16	14	6

Minerals (% Dry Weight)

K	S	Na	Ca	Mg
5	4	3	1	1

Trace Elements (mg/kg Dry Weight)

I	Fe	Zn	Cu
5000	100	30	3

Source: Data from Agrimer Algues Marines (2014).

Undaria pinnatifida (Harvey) Suringar, 1873

Phylum: Ochrophyta

Class: Phaeophyceae

Order: Laminariales

Family: Alariaceae

Common name: Wakame, Asian kelp, apron-ribbon vegetable

Distribution: Northwest Pacific Ocean; temperate Pacific Ocean

Habitat: Subtidal; often growing on manmade structures

Description: It has a branched holdfast giving rise to a stipe. Just above the holdfast, the stipe has very wavy edges, giving it a corrugated appearance. The stipe gives rise to a blade that is broad, flattened, and lanceolate. It has a distinct midrib. The margins of the blade are wavy. Plants can reach an overall length of 1–3 m.

Nutritional Facts

Proximate Composition

Crude protein (g/100 g dry weight)	16.3

Fiber (g/100 g Wet Weight)

Total Fiber	Soluble Fiber	Insoluble Fiber	Carbohydrate
3.4	2.9	0.5	4.6

Fiber and Carbohydrate (mg per 8 g)

Total Fiber	Soluble Fiber	Insoluble Fiber	Carbohydrate
2.8	2.4	0.4	3.9

Minerals (mg/100 g Wet Weight)

Ca	K	Mg	Na	Cu	Fe	I	Zn
112	62	79	449	0.2	3.9	3.9	0.3

Fatty Acids (% of Total Fatty Acid)

Saturated	Monounsaturated	Polyunsaturated
20.39	10.5	69.11

Source: Data from MacArtain et al. (2007).

Vitamins (mg per 8 g Dry Portion)

B_1	B_2	B_3	B_6	B_8	B_9	C	E	B_{12}
0.40	0.94	7.20	0.26	0.02	0.53	14.78	1.40	0.35

Source: Data from MacArtain et al. (2007).

Amino Acids (mg/g Dry Weight)

Aspartic acid	10.18
Glutamic acid	10.65
Serine	5.76
Threonine*	7.33
Glycine	8.76
Alanine	27.20
Arginine	8.41
Proline	5.52
Valine*	16.84
Methionine*	3.58
Isoleucine*	7.91
Leucine*	13.70
Tryptophan*	0.43
Phenylalanine*	7.80
Cystine	2.41
Lysine*	11.12
Histidine*	5.25
Tyrosine	4.31
Total	157.16

*Essential amino acids.

Vitamins (mg/100 g Dry Weight)

B_1	0.30
B_2	1.35
B_6	0.18
Niacin	2.56

Minerals (mg/100 g Dry Weight)

Ca	950
Mg	405
P	450
I	26
Na	6494
K	5691
Ni	0.265
Cr	0.072
Se	
Fe	1.54
Zn	0.944
Mn	0.332
Cu	0.185
Pb	0.079
Cd	0.028
Hg	0.022
As	0.055

Source: Data from Kolb et al. (2004).

RED ALGAE

Chondrus crispus Stackh

Phylum: Rhodophyta

Class: Rhodophyceae

Order: Gigartinales

Family: Gigartinaceae

Common name: Irish moss

Distribution: Ireland, Great Britain, Iceland and the Faroe Islands; western Baltic Sea to southern Spain; Atlantic coasts of Canada

Habitat: On rocks (above), in pools, and in lower intertidal and shallow subtidal areas

Description: This species has cartilaginous, dark purplish-red fronds. Female plants may be iridescent at the apices under water and greenish-yellow in upper-shore rock pools. It is 150 mm high. Stipe is compressed and narrow, and expands gradually onto a flat, repeatedly dichotomously branched blade. Axils are rounded, and apices are blunt or subacute. Frond is thicker in center than at margins.

Nutritional Facts

Proximate Composition

Dry matter (DM) (on wet weight)	19–25%
Carbohydrate (on DM)	50–60%
Proteins (on DM)	5–25%
Lipids (on DM)	0.6–6%

Minerals (mg/100 g Wet Weight)

Ca	K	Mg	Na	Cu	Fe	I	Zn
374	828	574	1573	0.1	6.6	6.1	NA

Source: Data from MacArtain et al. (2007).

Minerals (Average % in Dry Matter)

Sulfur	8.5
Sodium	6
Potassium	3.5
Calcium	1.5
Magnesium	1
Phosphorus	0.05

45

Trace Elements (Average in mg/kg Dry Matter)

Iodine	350
Iron	220
Boron	150
Nickel	6.5
Fluorine	6
Copper	4

Source: Data from Agrimer Algues Marines (2014).

Kappaphycus cottonii (Weber-van Bosse) Doty (= *Eucheuma cottonii*)

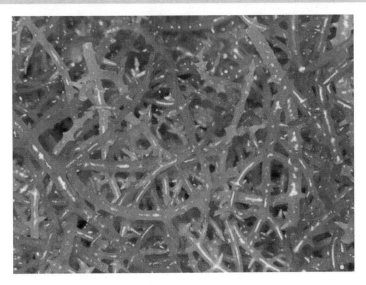

Phylum: Rhodophyta

Class: Rhodophyceae

Order: Gigartinales

Family: Solieriaceae

Common name: Guso

Distribution: Indo-Pacific region

Habitat: Rocky or solid corally substrates near the reef edge

Description: Plants consist of somewhat compressed irregular branches, attached to one another by undefined haptera forming slightly amorphous fronds. All branches have rough warty surfaces because of numerous short, blunt, and stubby spines or tubercles.

Nutritional Facts

Proximate Composition (%)

Protein	Lipid	Carbohydrate	Fiber	Ash	Moisture
9.76	1.10	26.49	5.91	46.19	10.55

Source: Data from Matunjan et al. (2009).

Eucheuma denticulatum (var. yellow) (Burman) Collins et Hervey (= *Eucheuma spinosum*)

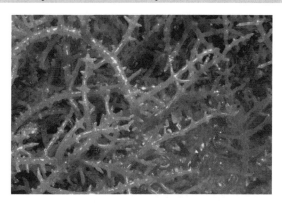

Phylum: Rhodophyta

Class: Rhodophyceae

Order: Gigartinales

Family: Solieriaceae

Common name: Spinosum

Distribution: Sri Lanka, Madagascar, Philippines

Habitat: Coarse sandy-corally to rocky substrates

Description: Thallus consists of many terete branches, tapering to acute tips. These are usually densely covered with 1–8 mm long spinose determinate branchlets arranged in whorls, forming distinct "nodes" and "internodes," especially at the terminal portion of the branches. It may form a dominant component of the algal community.

Nutritional Facts

Proximate Composition (% Dry Weight)

Ash	Carbohydrate	Lipid	Protein	Energy (kJ/g)
43.8	30.6	1.6	5.0	6.7

Source: Data from Ahmad et al. (2012).

Solieria robusta (Greville) Kylin

Phylum: Rhodophyta

Class: Rhodophyceae

Order: Gigartinales

Family: Solieriaceae

Common name: Blubber weed

Distribution: Western Australia to Tasmania

Habitat: Variety of depths and wave energies

Description: Plants of this species, which are 100–230 mm tall, are red, gray, or bleached yellow. They are fairly soft when fresh. A basal holdfast with finger-like, often orange extensions occurs.

Branches, which are 10–40 mm apart, are tubular (cylindrical) and pinched at the base, and they taper to a point.

Nutritional Facts

Proximate Composition

Protein (%)	Carbohydrate (%)	Lipid (%)
4.5	45	2.8

Mineral and Vitamin (ppm)

Calcium	Ascorbic Acid
7	27

Source: Data from Ambreen et al. (2012).

Spyridia filamentosa (Wulfen) Harvey

Phylum: Rhodophyta

Class: Florideophyceae

Order: Ceramiales

Family: Spyridiaceae

Common name: Hairy basket weed

Distribution: Indian Ocean, North Atlantic Ocean, North Sea, Red Sea, South Africa and Tanzania

Habitat: Eroded coral in sandy areas; sandy mud; shallow subtidal to 10 m

Description: Branches of this species are covered with many soft, fine, short branches. Plants are 2–18 cm and are fastened by small discoid holdfast. Lower portions of plants and lower branches tend to entangle. Branching is irregularly dichotomous or completely irregular and branches are covered with many soft, fine, short branches

Nutritional Facts

Proximate Composition and Polyunsaturated Fatty Acid (PUFA) (Wet Weight %)

Protein	Lipid	Moisture	Ash	PUFA
2.81	0.23	82.1	6.0	25.3

Source: Data from Polat and Ozogul (2013).

Hypnea pannosa J. Agardh, 1847

Phylum: Rhodophyta

Class: Florideophyceae

Order: Gigartinales

Family: Cystocloniaceae

Common name: Blue hypnea

Distribution: Southern Japan, China, Taiwan, the Philippines, Vietnam, Malaysia, Indonesia, Australia, Mauritius, Hawaii, the Indian Ocean, southern sea coast of Pakistan

Habitat: Rocks near the lower mark of the littoral zone; tide pools

Description: This species is distinguished by its cartilaginous, iridescent, compactly tufted fronds with densely interwoven branches. Thalli are a complex array of branches forming large, flattened clumps. Plant is red to purple in color and greenish when dried. Branching is subdichotomous to alternate in wide angles. Branchlets split into short, stubby spines.

49

Hypnea musciformis (Wulfen) J.V. Lamouroux 1813

Phylum: Rhodophyta

Class: Florideophyceae

Order: Gigartinales

Family: Cystocloniaceae

Common name: Crozier weeds

Distribution: Mediterranean, Philippines, Indian Ocean, and from the Caribbean to Uruguay

Habitat: Calm intertidal and shallow subtidal reef flats, in tide pools, and on rocky intertidal benches; epiphytic on *Sargassum* and other algae

Description: This species has clumps or masses of loosely intertwined, cylindrical branches. Branching is variable and irregular, often tendril-like, and twisted around axes of other algae. Ends of many axes and branches are flattened with broad hooks. Holdfasts are small, inconspicuous, or lacking. Plants are red or yellowish brown.

Hypnea japonica Tanaka 1941

Phylum: Rhodophyta

Class: Florideophyceae

Order: Gigartinales

Family: Cystocloniaceae

Common name: Japanese red algae

Distribution: North Taiwan and northeastern Taiwan.

Habitat: Intertidal coral reefs to subtidal zone 10 m deep

Description: Thalli are purple-red in color, fleshy, terete, erect, bushy, and entangled. They are irregularly branching with wide angles. Branches are 0.1–0.2 cm in diameter, and the tips of the branches are often elongate, swollen, and hooked.

Hypnea charoides Lamouroux

Phylum: Rhodophyta

Class: Florideophyceae

Order: Gigartinales

Family: Hypneaceae

Common name: Spiny red weed

Distribution: Native to Western Australia; South Africa; central west coast, Leeuwin-Naturaliste, Otway

Habitat: Under moderate wave action and epiphytic on the seagrass; amphibolis

Description: Plants are light red-brown or yellowish. They are 100–200 mm tall, with a tangled base and straggling upright branches. Tiny pointed spines are found all around branches.

Nutritional Facts

Proximate Composition (Dry Weight Basis)

	Hypnea pannosa	*Hypnea musciformis*
Protein (%)	16.31	18.64
Crude lipid (%)	1.56	1.27
Carbohydrate (%)	22.89	20.60
Fiber (%)	40.59	37.92
Ash (%)	18.65	21.57
Moisture (%)	12.35	11.54

Amino Acids (mg/g Protein)

	H. pannosa	*H. musciformis*	*H. japonica*	*H. charoides*
Arginine	51.9	68.3	66.8	63.6
Histidine	5.8	6.4	6.89	6.58
Isoleucine	48.2	42.1	44.8	48.5
Lysine	55.3	45.6	66.3	64.9
Leucine	77.4	84.8	97.9	72.3
Methionine	14.6	16.2	18.5	16.8
Phenylalanine	30.8	32.9	37.2	56.0
Tyrosine	25.8	26.2	27.9	26.0
Threonine	41.5	62.7	45.9	51.3
Valine	53.7	55.8	56.3	61.4
Alanine	52.4	47.9	57.4	52.3
Aspartic acid	76.3	86.5	98.4	88.6
Glutamic acid	118.8	107.2	110	98.4
Glycine	49.2	47.7	54.2	50.6
Proline	40.8	44.2	45.4	47.9
Serine	32.3	43.9	47.5	44.9
Total (g/100 g Dry Weight)	14.8	15.3	17.3	16.2

Source: Data from Abdul et al. (2013).

Kappaphycus alvarezii (Doty) Doty nov. comb (= *Eucheuma alvarezii*)

Phylum: Rhodophyta

Class: Florideophyceae

Order: Gigartinales

Family: Solieriaceae

Common name: NA

Distribution: Native to the Indo-Pacific; Western Pacific and Indian Oceans

Habitat: Just below the zero tide line to the upper subtidal portion of reef areas on sandy-corally to rocky substrate

Description: This species may be tall and loosely branched with few blunt or pointed determinate branchlets, or it may form the "spinosum" type where thallus may be densely branched and covered with coarse spinose branchlets. These, however, are not arranged in whorls so as to form distinct "nodes" and "internodes."

Nutritional Facts

Proximate Composition

	Moisture (% FS)	Carbohydrate (% DW)	Ash (% DW)	Protein (% DW)	Fiber (% DW)	Lipid (% DW)
Kappaphycus alvarezii (var. aring-aring)	79.78	66.66	23.25	5.35	4.50	0.23
Kappaphycus alvarezii (var. green tambalang)	79.65	62.50	26.25	5.63	5.45	0.18

FS: fresh sample; DW: dry weight.

Vitamins and Fatty Acids

Vitamin A (per 100 g)	865 µg
Unsaturated fatty acids	44.50% (of the total)
Oleic acid	11%
Cis-heptadecanoic acid	13.50%
Linoleic acid	2.3%
Saturated fatty acids	37.0%

Source: Data from Fayaz et al. (2005).

Minerals (mg per 100 g)

Calcium	1068
Phosphrous	124
Iron	0.93
Magnesium	152
Niacin	2.2

Kappaphycus striatum (Schmitz) Doty nov. comb. (= *Eucheuma striatum*)

Phylum: Rhodophyta

Class: Florideophyceae

Order: Gigartinales

Family: Solieriaceae

Common name: NA

Distribution: Hawaiian Islands

Habitat: Reef flat and reef edge, 1 to 17 m deep; loosely attached to broken corals

Description: Thallus is erect or decumbent. The axis is not percurrent, and branches are roughened by the presence of spinose processes or determinate branchlets. Axial hyphae are present in the branches and are about 5 mm in diameter.

Nutritional Facts

Proximate Composition

	Moisture (% Fresh Sample)	Carbohydrate (% Dry Weight)	Ash (% Dry Weight)	Protein (% Dry Weight)	Fiber (% Dry Weight)	Lipid (% Dry Weight)
Kappaphycus striatum var. *sacol* (var. katunai green)	79.70	66.02	22.99	5.42	5.34	0.22
Kappaphycus striatum var. *sacol* (var. katunai brown)	75.95	66.63	21.93	5.22	5.96	0.25
Kappaphycus striatum var. *sacol* (var. katunai yellow)	76.69	67.49	22.84	5.40	4.03	0.24

Source: Data from Ahmad et al. (2012).

Acanthopora spicifera (Vahl) Borgesen

Phylum: Rhodophyta

Class: Florideophyceae

Order: Ceramiales

Family: Rhodomelaceae

Common name: Spiny seaweed

Distribution: Tropics and subtropics

Habitat: Wide variety of substrata, from hard bottom to as an epiphyte on other algae, or as a free-living drift alga

Description: This species has a large, irregularly shaped holdfast for attachment to hard bottoms. From the holdfast, erect fronds begin to branch out. The main branches have short, determinate branchlets that are irregularly shaped and spinose. Branchlets are hook-like and brittle, and they fragment easily under heavy wave action. Color of the plant is red, purple, or brown. It grows upright to about 25 cm.

Nutritional Facts

Proximate Composition (%)

Protein	Sugar	Lipid
5.3	6.3	1.1

Minerals (µg/g)

Na	K	Ca	Cu	Fe	Mg	Mn	Zn
710	580	250	48.5	2602	8154	77.7	97.4

Source: Data from Seenivasan et al. (2012).

Amino Acids (% Dry Weight)

Aspartic acid	15.7
Alanine	5.5
Arginine	5.7
Glutamic acid	17.4
Glycine	4.3
Histidine	2.2
Isoleucine	3.6
Leucine	7.8
Lysine	8.6
Methionine	1.2
Phenylalanine	5.3
Proline	5.7
Serine	4.3
Threonine	5.9
Tyrosine	3.7
Valine	6.1

Source: Data from Rameshkumar et al. (2013).

Dasya rigidula (Kützing) Ardissone

Phylum: Rhodophyta

Class: Florideophyceae

Order: Ceramiales

Family: Dasyaceae

Common name: NA

Distribution: North Carolina, Bermuda, southern Florida, Gulf of Mexico, the Caribbean, the Mediterranean, from the British Isles to Portugal, the Azores, the Canary Islands, and West Africa

Habitat: Coastal waters

Description: The plants are bushy, pinkish to red in color, and 2 to 6 cm in height. Erect branches develop from a discoid base, giving rise to a subdichotomous to alternate branching pattern. Axes are cylindrical with five pericentral cells in each segment. The axis growth is sympodial.

Nutritional Facts

Proximate Composition and Polyunsaturated Fatty Acids (PUFA) (Wet Weight %)

Protein	Lipid	Moisture	Ash	PUFA
2.61	0.34	82.1	7.3	8.6

Source: Data from Polat and Ozogul (2013).

Gelidium pusillum (Stackhouse) Le Jolis

Phylum: Rhodophyta

Class: Florideophyceae

Order: Gelidiales

Family: Gelidiaceae

Common name: Dwarf gelidium

Distribution: Confined to the northeastern Atlantic

Habitat: Rocks in upper intertidal zones

Description: This species is cartilaginous, brownish-red or purplish-red when wet, and black or blackish-red when dry. It is a turf-forming species, 2–10 mm height. It is erect, and its reproductive fronds are flattened, leaflike, and 0.5–2 mm broad. Apices are spoon shaped.

Nutritional Facts

Proximate Composition (%)

Protein	Lipid	Carbohydrate	Fiber	Ash	Moisture
1.31	2.16	40.64	24.74	21.15	10.85

Amino Acids (mg/g Protein)

Essential Amino Acids (EAAs)	
Arginine	62.7
Histidine	4.6
Isoleucine	42.1
Lysine	48.3
Leucine	75.2
Methionine	15.8
Phenylalanine	31.9
Tyrosine	26.2
Threonine	51.5
Valine	44.7

Non-EAAs	
Alanine	51.6
Aspartic acid	82.2
Glutamic acid	108.8
Glycine	43.4
Proline	46.6
Serine	38.2
Total EAAs	403
Total amino acids (g/100 g dry weight)	9.8

Source: Data from Siddique et al. (2013b).

Laurencia papillosa (C. Agardh) Greville, 1830

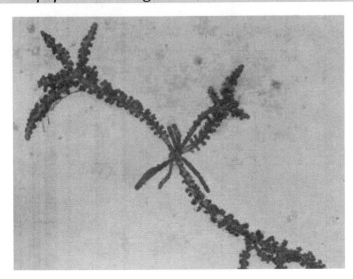

Phylum: Rhodophyta

Class: Florideophyceae

Order: Ceramiales

Family: Rhodomelaceae

Common name: NA

Distribution: Europe, India

Habitat: Sheltered waters

Description: Plants are 5–16 cm tall and grow in dense clusters. The lower parts of the plants are smooth, but toward the ends, the branches are sparsely to densely crowded with short, truncate to tuberculate branchlets. This species appears in shades of green, brownish-green, or crimson-green.

Nutritional Facts

Proximate Composition and Polyunsaturated Fatty Acids (PUFA) (Wet Weight %)

Protein	Lipid	Moisture	Ash	PUFA
0.80	0.26	88.7	3.7	27.3

Source: Data from Polat and Ozogul (2013).

Proximate Composition

	Moisture (% Fresh Sample)	Carbohydrate (% Dry Weight)	Ash (% Dry Weight)
Laurencia sp. (var. yellow)	96.03	66.78	8.90
Laurencia sp. (var. brown)	93.59	66.00	11.60

Proximate Composition (% Dry Weight)

	Protein	Fiber	Lipid
Laurencia sp. (var. yellow)	17.28	6.64	0.40
Laurencia sp. (var. brown)	14.80	7.16	0.45

Source: Data from Ahmad et al. (2012).

Pterocladia capillacea (Gmelin) Santelices & Hommersand

Phylum: Rhodophyta

Class: Florideophyceae

Order: Gelidiales

Family: Pterocladiaceae

Common name: Small agar weed

Distribution: Cosmopolitan: in southern Australia, from Venus Bay, South Australia, to Victoria

Habitat: Shallow water to 16 m deep on coasts with rough to moderate wave energy

Description: Plants are dark red and are 40–150 mm tall. They form loose turfs. The compressed main branches (axes) are flat branched. A tangled runner occurs at the base of upright axes.

Nutritional Facts

Minerals and Proximate Composition (mg/g)

Ca	Mg	Na	K	Fat	Carbohydrate
3.1	2.1	17.9	8.3	177.9	96.4

Source: Data from Ghada and El-Sikaily (2013).

Proximate Composition (% Dry Weight)

Protein	Carbohydrate	Lipid	Ash	Moisture
23.72	50.49	2.71	13.02	10.06

Amino Acids (% to Total Amino Acids)

EAA	Non-EAA	Total*
40.10	59.90	2836

EAA: essential amino acid.

*µg/g.

Fatty Acids (% to Total Fatty Acids)

SFA	MUFA	PUFA	Total*
70.00	14.90	15.10	514.78

SFA: saturated fatty acids; MUFA: monounsaturated fatty acids; PUFA: polyunsaturated fatty acids.

*µg/g.

Source: Data from Khairy and El-Shafay (2013).

59

Gracilaria cornea J. Agardh

Phylum: Rhodophyta

Class: Florideophyceae

Order: Gracilariales

Family: Gracilariaceae

Common name: Tropical red seaweed

Distribution: Asia, South America, Africa, and Oceania

Habitat: Attaches to rocks or dead corals

Description: This species forms patches (1–4 m in diameter) and grows at 2 to 4 m depth. The plants are about 0.5 m in size; they are usually bushy, and short and stubby to slender; long cartilaginous branches are seen. They are usually red to purple in color.

Nutritional Facts

Proximate Composition (%)

Protein	Lipid	Carbohydrate	Fiber	Ash	Moisture
5.47	—	36.29	5.21	29.06	—

Source: Data from Siddique et al. (2013a).

Gracilaria changgi B.M. Xia & I.A. Abbott

Phylum: Rhodophyta

Class: Florideophyceae

Order: Gracilariales

Family: Gracilariaceae

Common name: NA

Distribution: West coast of Peninsular Malaysia

Habitat: Mangrove swamps

Description: The plants have discoid holdfasts with thalli length ranging from 6 to 20 cm. Branching is irregular, alternate, or secund in two to four orders with an abrupt constriction at the base, forming a slender stipe. Distal end of stipe is slightly swollen, tapering toward the tip.

Nutritional Facts

Proximate Composition (%)

Protein	Lipids	Fiber	Ash	Vitamin C*
6.9	3.3	24.7	22.7	28.5

*mg/100 g.

Essential Amino Acids (g of Amino Acid/16 g of Nitrogen)

Isoleucine	5.7
Leucine	7.0
Lysine	3.5
Phenylalanine+ Tyrosine	9.5
Methionine+ Cysteine	5.8
Threonine	6.9
Valine	5.4
Arginine	9.0

Source: Data from Norziah and Ching (2000).

Minerals (mg/100 g Dry Basis except Cu in µg/100 g)

Ca	Zn	Fe	Cu
651	13.8	95.6	800

Source: Data from Ratana-arporn and Chirapart (2006).

Gracilaria compressa Grev.

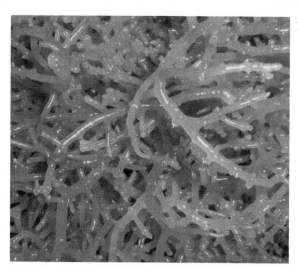

Phylum: Rhodophyta

Class: Florideophyceae

Order: Gracilariales

Family: Gracilariaceae

61

Common name: NA

Distribution: Mediterranean Sea—Eastern Basin

Habitat: Rock in the lower intertidal to sublittoral zones

Description: Frond is succulent, brittle, and somewhat compressed. It is alternately or subdichotomously branched. Branches are long and are mostly simple, tapering to a fine point. Tubercles are ovate or subglobose and sessile, and are scattered plentifully over the branches.

Nutritional Facts

Minerals and Proximate Composition (mg/g)

Ca	Mg	Na	K	Fat	Carbohydrate
3.3	1.7	29.1	4.5	41.0	116.2

Source: Data from Ghada and El-Sikaily (2013).

Gracilaria manilaensis Yamamoto & Trono

Phylum: Rhodophyta

Class: Florideophyceae

Order: Gracilariales

Family: Gracilariaceae

Common name: Filiform sea moss

Distribution: Pacific Ocean: Philippines

Habitat: Pebbles, shells, and stones on muddy bottoms in shallow, nutrient-rich water

Description: Thalli, which are 54 cm in length, are reddish brown to purple in color. They are erect, loosely branched, and attached to shells and debris by a very small discoid holdfast. Branches, which are 0.8 to 1.9 mm in diameter, are terete throughout; they are slender and gradually taper toward filiform apices. They are also distinctly constricted at the base. Nutritive filaments are abundant between the pericarp and gonimoblast.

Nutritional Facts

Proximate Composition

Moisture	
Fresh sample (WW%)	88.88
Dried sample (DW%)	6.08
Ash (DW%)	
Normal Dry (4 days at room temperature)	13.17
Freeze Dry	35.47
Protein (DW%)	
Normal Dry (4 days at room temperature)	10.77
Freeze Dry	9.77

Carbohydrate (DW%)	45.92
Lipid (DW%)	4.32
Fat (DW%)	0.920
Energy (cal/g)	3283

DW: dry weight.

Fatty Acids (%)

Total	25.47
MUFA	35.70
PUFA	42.18

MUFA: monounsaturated fatty acids; PUFA: polyunsaturated fatty acids.

Source: Data from Abdullah (2013).

Gracilaria salicornia (C. Agardh) Dawson 1954

Phylum: Rhodophyta

Class: Florideo**phyceae**

Order: Gracilariales

Family: Gracilariaceae

Common name: NA

Distribution: Warm Indian and Pacific Oceans

Habitat: Tidepools and on reef flats, intertidal to subtidal areas

Description: Thalli consist of solid, brittle, cylindrical to compressed branches, 2–5 mm in diameter. Axes are 3–18 cm long and 1.5 mm broad, with branches usually irregularly arranged. Both axes and branches are regularly or irregularly constricted or continuous. Plants are often prostrate and overlapping with lateral branches, or erect with an inconspicuous discoid holdfast.

Nutritional Facts

Proximate Composition (g/100 g Dry Weight)

Lipids	Protein	Fiber	Dry Weight	Ash
2.00	9.58	10.40	9.98	38.91

Amino Acids (mg/g Protein)

Aspartic acid	53.9
Glutamic acid	75.9
Serine	34.6
Glycine	75.6
Threonine	32.9
Arginine	75.8
Alanine	75.5
Tyrosine	75.9
Histidine	14.3
Valine	41.4
Methionine	77.5
Isoleucine	30.3
Phenylalanine	32.7
Leucine	76.6
Lysine	77.1
Proline	39.8
EAA	520.2
Non-EAA	369.6
Total	889.8

EAA: essential amino acid.

Fatty Acids (% of Total Fatty Acids)

Saturated	MUFA	PUFA
48.92	16.36	17.30

MUFA: monounsaturated fatty acids;
PUFA: polyunsaturated fatty acids.
Source: Data from Tabarsa et al. (2012).

Gracilaria verrucosa (Hudson) Paperfuss

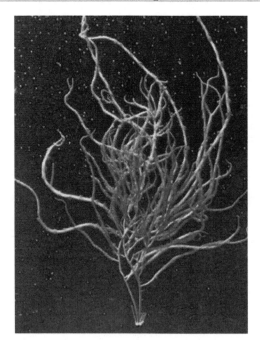

Phylum: Rhodophyta

Class: Florideophyceae

Order: Gracilariales

Family: Gracilariaceae

Common name: NA

Distribution: Philippines, India, Indonesia

Habitat: Mangrove swamps, brackish water

Description: Plants are bushy. Texture is firmly fleshy, and color is dull purplish, grayish, or greenish translucent. Branches are 0.5–2.0 mm in diameter and are repeatedly dividing alternately or occasionally dichotomously branched with numerous lateral proliferations.

Nutritional Facts

Proximate Composition

Moisture (% Fresh Sample)	Carbohydrate (% Dry Weight)	Ash (% Dry Weight)
85.45	74.11	6.05

Proximate Composition (% Dry Weight)

Protein	Fiber	Lipid
11.73	7.84	0.27

Source: Data from Ahmad et al. (2012).

Palmaria palmata (Linnaeus) Weber & Mohr

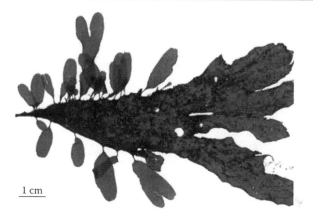

1 cm

Phylum: Rhodophyta

Class: Florideophyceae

Order: Palmariales

Family: Palmariaceae

Common name: Red dulse

Distribution: Northern coasts of the Atlantic and Pacific oceans; western Pacific

Habitat: On rock, mussels, and epiphytic on several algae; intertidal and shallow subtidal

Description: This species is reddish-brown, membranous, or leathery. Fronds are flattened and usually between 20 and 50 cm in length, but sometimes up to 1 m and are arising from a discoid base. It usually has a small stipe expanding gradually to form simple or dichotomously and palmately divided fronds, often with characteristic marginal leaflets. Blade is very variable in shape with broadly ovate to narrowly linear segments.

Nutritional Facts

Proximate Composition
Fiber (g/100 g Wet Weight)

Total Fiber	Soluble Fiber	Insoluble Fiber	Carbohydrate
5.4	3.0	2.3	10.6

Fiber and Carbohydrate (mg per 8 g)

Total Fiber	Soluble Fiber	Insoluble Fiber	Carbohydrate
2.7	1.5	1.2	5.3

Minerals (mg/100 g Wet Weight)

Ca	K	Mg	Na	Cu	Fe	I	Zn
149	1170	98	255	0.4	12.8	10.2	0.3

Polyunsaturated Fatty Acid (PUFAs) (% of Total Fatty Acid)

Saturated	Monounsaturated	Polyunsaturated
60.48	10.67	28.86

Vitamins (mg per 8 g Dry Portion)

B_1	B_2	B_3	B_6	B_8	B_9	C	E	B_{12}
0.02	0.08	0.80	0.002	0.002	0.02	5.52	1.30	1.84

Source: Data from MacArtain et al. (2007).

Jania rubens (Linnaeus) Lamouroux

Phylum: Rhodophyta

Class: Florideophyceae

Order: Corallinales

Family: Corallinaceae

Common name: NA

Distribution: Baltic Sea; Mediterranean; Indian Ocean, Black Sea, and China Sea

Habitat: Lower intertidal regions; epiphytic on older plants of the brown algae

Description: This species is slender, rose-pink, and articulated. Calcified fronds are seen in rounded bunches that are 50 mm high. The plant is repeatedly dichotomously branched. Segments are cylindrical and are 100 µm in diameter.

Nutritional Facts

Proximate Composition and Polyunsaturated Fatty Acid (PUFA) (Wet Weight %)

Protein	Lipid	Moisture	Ash	PUFA
1.66	0.33	42.8	46.1	41.4

Source: Data from Polat and Ozogul (2013).

Porphyra tenera Kjellman, 1897

Phylum: Rhodophyta
Class: Florideophyceae
Order: Bangiales
Family: Bangiaceae
Common name: Asakusa nori
Distribution: Northwest Pacific: Japan, China; Indian Ocean: Mauritius
Habitat: Intertidal zone, typically between the upper intertidal zone and the splash zone; sometimes settled on molluscs and other seaweed species

Description: Thallus is small, irregularly shaped, folliaceus (leaf-like), and membranaceous. It is attached to the substrate (e.g., rocks) by a small discoid holdfast. Fronds are more or less divided, are crinkled, and undulate at the edges. The color of the plant is green in its earlier stages, becoming brownish-purple or purplish-red at older stages.

Porphyra haitanensis T. J. Chang et B. F. Cheng

67

Phylum: Rhodophyta

Class: Florideophyceae

Order: Bangiales

Family: Bangiaceae

Common name: NA

Distribution: China

Habitat: Lower intertidal and subtidal habitats

Description: Thalli are over 28 cm in height. Blades are lanceolate, subovate, or elongated subovate, and are conspicuously stiped. Marginal portions are slightly undulate and membranaceous. This species has a single, axial, stellate chromatophore normally, but sometimes it has two chromatophores.

Nutritional Facts

Proximate Composition (%)

	P. tenera	P. haitanensis
Ash	9.07	8.78
Crude lipid	2.25	1.96
Crude protein	36.88	32.16

Amino Acids (mg/100 g)

	P. tenera	P. haitanensis
Taurine	979.04	646.55
Aspartic acid	141.98	171.37
Threonine	31.80	86.43
Serine	20.02	44.81
Asparagine	22.37	86.55
Glutamic acid	843.35	277.45
Glycine	22.06	26.11
Alanine	936.28	1,218.71
Citrulline	77.80	71.32
Valine	33.48	—
Isoleucine	46.67	49.88
Leucine	27.92	33.22
γ-amino butyric acid	31.34	—

Source: Data from Hwang et al. (2013).

	P. tenera	P. haitanensis
Moisture	3.66	6.74

Porphyra umbilicalis Kützing

Phylum: Rhodophyta

Class: Florideophyceae

Order: Bangiales

Family: Bangiaceae

Common name: Purple laver

Distribution: North Atlantic; Iceland; from Norway to Portugal and Western Mediterranean; from Labrador in Canada to the mid-Atlantic coast of the United States

Habitat: Generally on mussels, sometimes on rock; midtidal to splash zone

Description: This species is usually 5 to 10 cm tall, and up to 20 cm across. It has an irregularly shaped, broad frond that is membranous but tough. The blade expands around the holdfast and appears pleated, or it resembles a rosette or a little cabbage. It exhibits a variety of colors and hues. The blades are greenish when young, then become purplish-red and light to dark reddish-brown.

Nutritional Facts

Proximate Composition
Fiber (g/100 g Wet Weight)

Total Fiber	Soluble Fiber	Insoluble Fiber	Carbohydrate
3.8	3.0	1.0	5.4

Fiber and Carbohydrate (mg per 8 g)

Total Fiber	Soluble Fiber	Insoluble Fiber	Carbohydrate
2.7	2.1	0.7	3.8

Minerals (mg/100 g Wet Weight)

Ca	K	Mg	Na	Cu	Fe	I	Zn
34	302	108	120	0.1	5.2	1.3	0.7

Vitamins (mg per 8 g Dry Portion)

B_1	B_2	B_3	B_6	B_8	B_9	C	E	B_{12}
0.08	0.27	0.76	0.12	NA	1.00	12.89	0.11	0.77

Source: Data from MacArtain et al. (2007).

HALOPHYTE

Salicornia bigelovii Torr.

Phylum: Tracheophyta

Class: Magnoliopsida

Order: Caryophyllales

Family: Chenopodiaceae

Common name: Dwarf saltwort, dwarf glasswort

Distribution: Native to coastal areas of the eastern and southern United States, southern California, and coastal Mexico

Habitat: Salt marshes; halophyte

Description: This is an annual herb producing an erect, branching stem that is jointed at many internodes. The fleshy, green to red stem may reach about 60 cm in height. The leaves are usually small plates, pairs of which are fused into a band around the stem. The inflorescence is a dense, sticklike spike of flowers. Each flower is made up of a fused pocket of sepals enclosing the stamens and stigmas, with no petals. The fruit is a utricle containing tiny, fuzzy seeds.

Nutritional Facts

Proximate Composition (g/100 g Fresh Weight)

Moisture	Protein	Lipids	Fiber	Carbohydrate	Ash
88.42	1.54	0.37	0.83	4.48	4.36

Biochemical Composition (mg/kg)

Chlorophyll	B-Carotene	Ascorbic Acid
569.1	159.0	58.4

Minerals (mg/g Fresh Weight)

Na	Zn	K	Cu	Mg	Ca	P	Fe
9.98	4.05	1.76	0.91	1.18	0.62	0.18	0.01

Amino Acids (g/kg Fresh Weight)

Asparagine	1.16
Cysteine	0.03
Valine	0.59
Serine	0.68
Methionine	0.09
Histidine	0.26
Phenylalanine	0.55
Glycine	0.53
Isoleucine	0.47
Threonine	0.55
Leucine	0.94
Arginine	0.68
Lysine	0.73
Alanine	0.69
Proline	0.83
Tyrosine	0.44
Total	10.86

Source: Data from Lua et al. (2010).

4
JELLYFISH

Rhopilema esculentum **Kishinouye, 1891**

Phylum: Cnidaria

Class: Scyphozoa

Order: Rhizostomae

Family: Rhizostomatidae

Common name: Japanese edible jellyfish

Distribution: Indo-West Pacific

Habitat: Coastal waters

Description: Umbrella of this species is about half as high as broad when swimming. It is rigid and thick in the central part while thin in the margin. When flattened, it has a maximum diameter of 70 cm. The exumbrella is smooth, and long and short marginal grooves are seen alternately. Eight rhopalia are located in the umbrella margin. A ring canal runs to connect the midpoints

71

of the radial canals. Gonads are belt shaped and folded complexly on the lower stomach wall. Mouth-arms are fused with each other for the proximal one-fourth of their length.

Nutritional Facts

Proximate Composition (g/100 g)

Energy	Protein	Fat	Carbohydrate
33 Kcal	3.70	0.30	3.80

Vitamins (mg/100 g)

Thiamine	Riboflavin	Niacin	E2
0.03	0.05	0.20	13

Minerals (mg/100 g)

Ca	K	Na	Mg	Fe	Zn	Se *	Cu	Mn
150	160	325	124	4.0	0.6	15.5	0.1	0.4

Micrograms.

Fats and Fatty Acids (per 100 g)

Fat	1.40 g
Saturated fatty acids	0.273 g
Hexadecanoic acid	0.190 g
Octadecanoic acid	0.063 g
Tetradecanoic acid	0.019 g
Monounsaturated fatty acids	0.202 g
Polyunsaturated fatty acids	0.475 g

Source: Data from Xiguang et al. (2004).

Stomolophus meleagris (Agassiz, 1860)

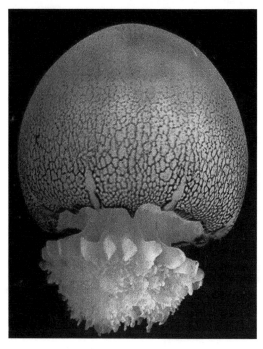

Order: Rhizostomae

Family: Stomolophidae

Common name: Cannonball jelly, jellyballs

Distribution: Western Atlantic; eastern Pacific; western Pacific

Habitat: Estuarine and saline waters

Description: The hemispherical bell of this species is 20–25 cm in diameter, and it has the general appearance of a large, half-egg-shaped mushroom. The color is milky bluish or yellowish, showing brown reticulations over its entire surface. The margin is densely pigmented brown with distinct spots. Eight short mouth-arms are joined together to form a stem-like mouth tube.

Nutritional Facts

Calories	30
Calories from fat	0
Total fat	0 g
Saturated fat	0 g
Cholesterol	0 mg
Sodium	20 mg
Total carbohydrates	0 g
Protein	8 g

Source: Data from Florida Department of Agriculture and Consumer Services and FoodReference.com.

5
CRUSTACEANS

SHRIMPS

Pandalus jordani **Rathbun, 1902**

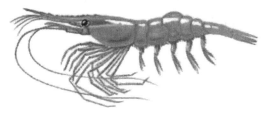

Phylum: Arthropoda

Class: Malacostraca

Order: Decapoda

Family: Pandalidae

Common name: Pacific pink shrimp, ocean shrimp

Distribution: Tropical to temperate; Eastern Pacific; Alaska, Canada, and USA

Habitat: On or near the seabed; depth range 36–457 m

Description: This species, which grows to a maximum length of 3 cm, is usually found on green mud or mixed mud and sand bottoms. Species is protandric hermaphrodite. Only one brood is produced every year.

Pandalus borealis **(Krøyer, 1838)**

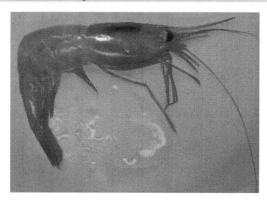

Order: Decapoda

Family: Pandalidae

Common name: Pink shrimp, deep-water prawn

Distribution: Atlantic and Pacific coastal waters

Habitat: Soft muddy bottoms; depths of 20–1330 m

Description: It is a medium-sized shrimp with a rather slender body. It has a uniform pink color with no banding. The rostrum is 1.5 to 2 times longer than the carapace. The third abdominal

segment, at the bend in the abdomen, has a distinctive dorsal spine. There are also single spines on the rear margin of the third and fourth abdominal segments. Maximum total length is 120 mm (male) and 165 mm (female).

Nutritional Facts

Proximate Composition

	Moisture (%)	Protein (%)	Fat (%)	Ash (%)
P. jordani	78.6	18.8	1.6	1.9
P. borealis	80.1	18.1	1.0	1.3

Source: Data from Krzynowek and Murphy (1987).

Fenneropenaeus penicillatus (Alcock, 1905)

Order: Decapoda

Family: Penaeidae

Common name: Redtail prawn

Distribution: Indo-West Pacific: from Pakistan to Indonesia

Habitat: Demersal; depth range 2–90 m

Description: The red tail is very similar to *P. indicus*. Like the Indian white, colour of this species varies from almost transparent to yellowish. The maximum total length and carapace length are 21.2 cm and 3.1 cm, respectively.

Nutritional Facts

Proximate Composition (% Wet Weight)

Lipid	Protein	Carbohydrate	Ash	Dry matter	Moisture
1.50	19.00	0.66	1.86	23.00	77.00

Source: Data from Nisa and Sultana (2010).

Parapenaeus longirostris (Lucas, 1847)

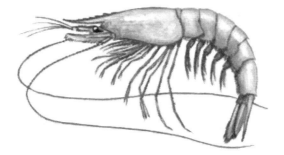

Order: Decapoda

Family: Penaeidae

Common name: Deep seawater rose shrimp

Distribution: Eastern Atlantic, Western Atlantic, and Mediterranean

Habitat: Sandy-muddy bottoms between 100 and 400 m

Description: It has a pink-orange carapace with a reddish rostrum. On the carapace, there is a long furrow beginning near the eyes and present on the entire length of the carapace. The telson ends with three sharp, hard little teeth. This species can grow up to 16 cm (males) and 19 cm (females) in total length.

Plesionika martia (Milne-Edwards, 1883)

Order: Decapoda

Family: Pandalidae

Common name: Golden shrimp

Distribution: Atlantic and the Mediterranean

Habitat: Demersal; depth range 176–700 m

Description: This species has a maximum length and weight of 17 cm and 5 g, respectively. It feeds on crustaceans (e.g., Pasyphaeidae and euphausiids) and carrion. Seasonal size distributions showed a size predominance of females over males. Ovigerous females are observed throughout the year.

Nutritional Facts

Proximate Composition (%)

	Moisture	Protein	Lipid	Ash
P. longirostris	78.7	20.0	1.1	1.6
P. martia	82.2	14.2	2.6	1.01

Fatty Acids (%)

	P. longirostris	*P. martia*
Total SFA	31.8	27.5
Total MUFA	26.1	34.5
Total PUFA	42.1	35.0

SFA: saturated fatty acids; MUFA: mono-unsaturated fatty acids; PUFA: polyunsaturated fatty acids.

Minerals (mg/kg)

	Mn	Cu	Zn	Fe	Mg	Ca	Na	P	K
P. longirostris	0.729	2.2	6.1	18	382	495	876	933	996
P. martia	0.145	2.83	5.9	2.0	579	322	574	1344	644

Source: Data from Öksüz et al. (2009).

Litopenaeus vannamei (Boone, 1931)

Order: Decapoda

Family: Penaeidae

Common name: White leg shrimp

Distribution: Eastern Pacific Ocean

Habitat: Ocean, at depths of up to 72 m; juveniles in estuaries

Description: It grows to a maximum length of 230 mm with a carapace length of 90 mm. The rostrum is moderately long, with 7–10 teeth on the dorsal side and 2–4 teeth on the ventral side. Coloration is normally translucent white, but it can change depending on substratum, feed, and water turbidity.

Nutritional Facts

Proximate Composition (%)

Protein	Carbohydrate	Lipid	Moisture	Ash
35.7	3.2	19.0	76.2	1.2

Source: Data from Nisa and Sultana (2010).

Minerals (mg/g)

Ca	Mg	Na	K	P	Mn	Fe
155	13.4	67.7	56.7	7.0	0.9	4.5

Source: Data from Ghada and El-Sikaily (2013).

Essential Amino Acids (%)

Arginine	1.2
Histidine	1.08
Isoleucine	12.3
Leucine	5.63
Lysine	13.42
Methionine	13.06
Phenylalanine	1.27
Tryptophan	1.3
Valine	23.72
Total	72.98

Nonessential Amino Acids (%)

Alanine	1
Asparagine	0.056
Aspartic acid	1.46

Cystine	5.56
Glutamic acid	2.51
Glycine	9.8
Proline	4.26
Serine	2.66
Tyrosine	2.51
Total	29.816

Fatty Acids (%)

SFA	11.1
MUFA	12.5
PUFA	14.3

SFA: saturated fatty acids; MUFA: monoun-saturated fatty acids; PUFA: polyunsaturated fatty acids.

Source: Data from Gunalan et al. (2013).

Penaeus monodon Fabricius, 1798

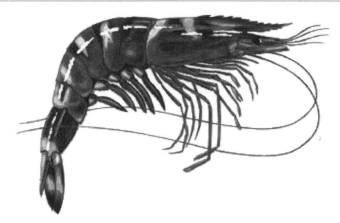

Order: Decapoda

Family: Penaeidae

Common name: Giant tiger prawn, Asian tiger shrimp

Distribution: Indo-Pacific

Habitat: Mud or sand bottoms at all depths from shallows to 110 m (offshore and inshore areas)

Description: It has rusty brown color and distinctive black-and-white banding across its back and tail. Females of this species can reach approximately 33 cm long with a weight of 200–320 g. Males are slightly smaller at 20–25 cm long and weigh 100–170 g.

Nutritional Facts

Proximate Composition (%)

Protein	Carbohydrate	Moisture	Lipid	Ash
36.14	2.11	75.18	1.40	3.8

Source: Data from Pushparajan et al. (2012).

Litopenaeus stylirostris Stimpson, 1874

Order: Decapoda

Family: Penaeidae

Common name: Blue shrimp

Distribution: Eastern Pacific: from Baja California (Mexico) to Peru

Habitat: Bottom mud and clay or sandy mud. Marine (adults) and estuarine (juveniles). Depth 0 to 27 m

Description: The longevity of this species has been estimated as 1.7 years. Size at first maturity is 109 mm (abdominal length). There are interannual variations in reproduction, as well as in the recruitment pattern. It grows to a maximum total length of 230 mm with a maximum carapace length of 59 mm.

Nutritional Facts

Proximate Composition (g/100 g)

Moisture	Protein	Lipid	Ash
73.0	20.5	1.0	2.3

Source: Data from Puga-López et al. (2013 a,b).

Acetes japonicus Kishinouye, 1905

Order: Decapoda

Family: Sergestidae

Common name: Akiami paste shrimp

Distribution: Indo-West Pacific, west coast of India to Korea, Japan, China, and Indonesia

Habitat: Shallow coastal waters over a muddy bottom

Description: This epipelagic species has a maximum length of 2.4 cm (male) and 3 cm (female). It is characterized by the loss of the fourth and fifth pairs of pereiopods. This small prawn is translucent, but with a pair of black eyes and a number of red pigment spots on the uropods.

Nutritional Facts

Proximate Composition (g/20 g)

Protein	Fat	Carbohydrate
3.0	0.64	0.04

Vitamins (mg/20 g)

E	B_1	B_2	B_3	B_6	B_5	C	A*	B_{12}*	B_9
0.5	0.03	0.05	0.38	0.02	0.1	0.4	36	1.24	9.8

Micrograms.

Minerals (mg/20 g)

Na	K	Ca	Mg	P	Fe	Zn	Cu	Mn
84	64	72	17	62	0.16	0.2	0.46	0.03

Source: Data from CalorieSlism (2014).

Metapenaeus affinis (Milne-Edwards, 1837)

Order: Decapoda

Family: Penaeidae

Common name: Jinga shrimp

Distribution: Indo-Pacific

Habitat: Demersal; brackish; depth range 5–92 m

Description: Males and females have been found to attain lengths of 145 mm and 174 mm, respectively, at the end of one year, and their life spans are 1.16 and 1.4 years, respectively.

Nutritional Facts

Proximate Composition (%)

Protein	Lipid	Carbohydrate	Ash	Moisture	Energy*
14.11	2.183	9.067	5.76	68.88	137.35

*cal/gDW.

Source: Data from Abdul-Sahib and Ajeel (2005).

Penaeus indicus (H. Milne Edwards, 1837)

Order: Decapoda

Family: Penaeidae

Common name: Indian white prawn

Distribution: Indo-West Pacific

Habitat: Bottom mud or sand; depth range 2–90 m; adults are marine and juveniles are estuarine.

Description: It has a maximum total length of 184 mm (male) and 228 mm (female). Its maximum carapace length is 56 mm. The post larvae of this species migrate to the estuaries, feed, and grow until they attain a length of 110–120 mm, whereupon these subadults return to the sea and get recruited into fishery.

Nutritional Facts

Proximate Composition (%)

	Carbohydrate	Protein	Lipid
Male	1.89	42.88	8.57
Female	1.91	40.68	8.92

Vitamins (mg/100 g Protein)

	B$_1$	B$_2$
Male	0.26	0.32
Female	0.32	0.42

Essential Amino Acids (mg/100 g)

	Male	Female
Arginine	4.28	4.97
Histidine	1.39	1.30
Lysine	1.50	1.34
Threonine	1.75	1.48
Methionine	1.28	1.46
Leucine	2.57	2.52
Isoleucine	1.77	1.50
Valine	2.30	1.97
Phenyl alanine	2.58	2.26

Nonessential Amino Acids (mg/100 g)

	Male	Female
Proline	2.97	2.81
Tyrosine	2.09	2.44
Glycine	2.52	2.73
Alanine	1.94	1.68
Serine	2.19	2.28
Glutamic acid	5.42	4.71
Aspartic acid	4.45	4.17

Minerals (mg/100 g Dry Weight)

	Na	K	Zn	Ca	P	S	Se
Male	31.65	33.27	40.62	45.09	74.32	85.43	0.02
Female	29.68	29.89	43. 17	39.96	75.45	102.47	0.03

Source: Data from Salam (2013).

CRABS

Portunus pelagicus (Linnaeus, 1758)

Order: Decapoda

Family: Portunidae

Common name: Flower crab, blue crab

Distribution: Throughout the Indian and West Pacific Oceans

Habitat: Sandy and sand-muddy depths in shallow waters 10–50 m in depth; near reefs, mangroves, seagrass, and algal beds

Description: Carapace, which is 20 cm wide, is rough to granulose with regions discernible. Front is with four acutely triangular teeth. Nine teeth are present on each anterolateral margin. The most external tooth is 2–4 times larger than the precedent. Chelae are elongate (more in males than in females) with conical teeth at the base of fingers. Legs are laterally flattened to varying degrees. The last two segments of the last pair are paddle-like. Males are colored with blue markings, and females are dull green.

Nutritional Facts

Proximate Composition

Protein (%)	Lipid (%)	Moisture (%)
23.23	1.19	73.60

Minerals (µg/g)

Cu	Zn	Fe
26.9	107.3	15.2

Source: Data from Ayas and Özoğul (2011).

Portunus sanguinolentus (Herbst, 1783)

Order: Decapoda

Family: Portunidae

Common name: Blood-spotted swimming crab

Distribution: Indo-Pacific

Habitat: Marine from littoral line to 30 m deep; sand, mud, or broken shelly substrata

Description: Carapace of this species is very broad and slightly convex. Front is cut into four sharp and very distinct teeth. The antero-lateral borders are very long and oblique and are cut into nine teeth, the last of which is about four times as long as any of the others. Chelipeds in the adult male are about 2 2/3 times the length of the carapace, but they are rather less in the female and young male. Arm is with three or four large spines on the anterior border.

Nutritional Facts

Proximate Composition (Hard and Soft Shell Crabs)

Crab	Protein (%)	Carbohydrate (%)	Lipid (%)
Hard shell	32.6	1.17	2.41
Soft shell	17.17	0.68	1.50

Essential Amino Acids

Amino Acids	Hard Shell (%)	Soft Shell (%)
Threonine	4.884	5.259
Valine	6.484	6.953
Arginine	8.385	9.242
Isoleucine	5.407	3.137
Leucine	8.386	9.242
Lysine	6.963	6.660
Phenylalanine	6.145	ND
Histidine	4.442	3.134

ND: no data.

Nonessential Amino Acids

Amino Acids	Hard Shell (%)	Soft Shell (%)
Glutamic acid	11.539	12.383
Tyrosine	1.913	2.054
Taurine	4.884	ND
Alanine	2.936	3.137
Aspargine	12.877	12.295
Glycine	4.442	4.799
Proline	5.544	4.681
Serine	9.648	10.370

ND: no data.

Minerals (mg/100 g)

Minerals	Hard Shell	Soft Shell
Calcium	2.028	1.022
Sodium	0.521	0.520
Potassium	0.480	0.497
Zinc	0.445	0.457
Magnesium	0.511	0.522

Source: Data from Sudhakar et al. (2009).

84

Scylla serrata Forsskål, 1775

Order: Decapoda

Family: Portunidae

Common name: Giant mud crab, mangrove crab

Distribution: Indo-Pacific

Habitat: Muddy bottoms, mangrove marshes, and river mouths in estuarine environments

Description: The carapace of this species has four blunt frontal teeth and each anterolateral margin has nine similarly sized broad teeth. The chilipeds (claws) are robust with several well-developed spines and the rear legs are flattened into swimming appendages. Individuals are grayish-green to purple-brown and variable in color with small irregular white spots on the carapace and swimming legs.

Nutritional Facts

Proximate Composition

Moisture (%)	Protein (%)	Fat (%)	Ash (%)	Carbohydrate (%)	Energy (Cal/100 g)
79.2	17.5	0.2	1.4	2.7	82.7

Source: Data from DailyBurn Tracker (2014).

Callinectes sapidus Rathbun, 1896

Order: Decapoda

Family: Portunidae

Common name: Blue Crab

Distribution: Western Atlantic Ocean from Nova Scotia to Argentina; Asia and Europe

Habitat: Tidal creeks

Description: It is easily identified by its body color, which is generally a bright blue along the frontal area, especially along the chelipeds. The rest of the body is shaded an olive-brown color. As with other Portunids, the fifth legs are adapted to a paddle-like shape to accommodate swimming. Females are identified due to their triangular or rounded aprons and red fingers on the chela.

Nutritional Facts

Proximate Composition of Carapace Meat (%)

	Protein	Lipid	Moisture	TMS*
Female	22.45	0.96	74.30	2.00
Male	21.40	1.11	75.47	1.85

* TMS –Total mineral substance

Fatty Acids (% of total fatty acids)

	SFA	MUFA	PUFA	Unidentified
Female	24.76	29.57	39.15	6.52
Male	23.27	26.63	42.80	7.30

Minerals (µg/g)

	Cu	Zn	Fe
Female	29.48	127.84	24.10
Male	21.55	113.35	22.61µ

Source: Data from Ayaz and Ozogul (2011).

Cancer magister (Dana, 1852)

Order: Decapoda

Family: Cancridae

Common name: Dungeness crab

Distribution: From Alaska's Aleutian Islands to Point Conception, California; Atlantic Ocean

Habitat: Eelgrass beds and water bottoms

Description: The carapace width of mature crabs may reach 25 cm. Dungeness crabs have a wide, long, hard shell, which they must periodically molt to grow. They have five pairs of legs, which are similarly armored, the foremost pair of which ends in claws that the crab uses both as defense and to tear apart large food items.

Nutritional Facts (per 85 g)

Proximate Composition (g)

Protein	Carbohydrate	Lipid	Ash	Moisture
14.8	0.63	0.82	1.45	67.3

Minerals (mg)

Ca	Fe	Mg	P	K	Na	Zn	Cu*	Mn
39.1	0.3	38.3	155	301	251	3.6	0.6	0.07

*µg

Fat and Fatty Acids

Total fat	0.529 g
Saturated	0.11 gg
Monounsaturated	0.14 g
Polyunsaturated	0.27 g
Cholesterol	50.15 g

Vitamins

C (total ascorbic acid)	2.98 mg
Thiamin	0.04 mg
Riboflavin	0.14 mg
Niacin	2.67 mg
B_6	0.13 mg
Folate	37.4 µg
B_{12}	7.65 µg
A (retinol activity equivalent)	22.95 µg
A	76.15 IU
Retinol	22.95 µg
Pantothenic acid	0.3 mg

Paralithodes camtschatica (Tilesius, 1815)

Order: Decapoda

Family: Lithodidae

Common name: Red king crab

Distribution: Native to the Bering Sea, north Pacific Ocean, Soviet Union, Murmansk Fjord, and Barents Sea

Habitat: Sandy and muddy areas in the intertidal zones

Description: This species has a carapace width of 28 cm and a leg span of 1.8 m. It has been named after the color it turns when it is cooked rather than the color of a living animal, which tends to be more burgundy.

Nutritional Facts (per 133 g, Cooked)

Calories	130
Protein	26 g
Total fat	2.1 g
Selenium	53.6 µg
Magnesium	84.4 mg
Vitamin B_6	0.24 mg
Folate	68 µg

Source: Data from All-Fish-Seafood-Recipes.com (2014).

87

Chionoecetes opilio (O. Fabricius, 1788)

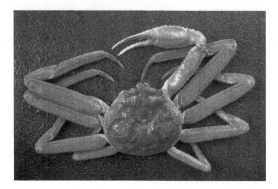

Order: Decapoda

Family: Oregoniidae

Common name: Snow crab

Distribution: Atlantic Ocean, North Pacific area from the Bering Strait to the Aleutian Islands, Japan, and Korea

Habitat: Ocean's benthic shelf and upper slope, in the sandy and muddy bottoms, and in depths from 20 to 1200 m

Description: Snow crabs, which measure 90 mm in length, have equally long and wide carapaces. They have a horizontal rostrum and triangular spines. Their first three walking legs are compressed, and their chelipeds are usually smaller, shorter, or equal to their walking legs. Snow crab are iridescent and range in color from brown to light red on top and from yellow to white on the bottom, and they are bright white on the sides of their feet.

Nutritional Facts (per 100 g)

Calories	90
Calories from fat	11
Total fat	1.2 g
Saturated fat	0.1 g
Cholesterol	55 mg
Sodium	539 mg
Potassium	173 mg
Protein	18.5 g

Source: Data from Calorie Count (2014).

Geryon quinquedens Smith, 1879

Order: Decapoda

Family: Geryonidae

Common name: Red deepsea crab

Distribution: Gulf of Maine, Gulf of Mexico, off southern New England

Habitat: Mid- to upper continental slope (>40 m)

Description: This species has a maximum size of about 150 mm carapace width (CW), and females are slightly smaller (140 mm CW). It feeds on rotifers, brine shrimp, and chopped mollusk meats. It may also feed on scavenge deadfalls (e.g., trawl discards) of fish and squid.

Pleuroncodes planipes Stimpson, 1860

Order: Decapoda

Family: Munididae

Common name: Pelagic red crab

Distribution: Mexico; San Diego, California; and Chile

Habitat: Continental shelf

Description: It is a bright red animal, up to 13 cm long. It has a shorter abdomen. It swims backwards by flipping its tail and streamlining its legs. It usually feeds on diatoms and zooplankton. It spends the majority of the year hiding in and around sandy bottoms.

Chionoecetes bairdi Rathbun, 1924

Order: Decapoda

Family: Oregoniidae

Common name: Tanner crab

Distribution: Japan: Funka Bay, off Noboribetsu, off Hiroo, off Kitami, off Hirono, and off Kushiro, Hokkaido; Bering Sea; Bristol Bay, Aleutian Islands; Alaska; British Columbia

Habitat: Subtidal, sandy, or muddy bottoms

Description: The branchial regions of this species are more depressed. The width of the animal exceeds the length. There are no dorsal spines. In this species, the last (posterior) three or four spines of the pterygostomial-branchial row are notably enlarged. The whole animal is rough. The carapace is narrower across the orbits; the outer orbital teeth are bent more inward. Chelipeds and legs are more coarsely and abundantly spinous.

Nutritional Facts

Proximate Composition

Species	Moisture (%)	Protein (%)	Fat (%)	Ash (%)
Geryon quinquedens	80.9	15.1	0.9	1.8
Pleuroncodes planipes	80.0	8.2	1.6	4.0
Chionoecetes bairdi	80.0	18.8	1.6	0.4

Source: Data from Krzynowek and Murphy (1987).

LOBSTERS

Panulirus interruptus (J.W. Randall, 1840)

Order: Decapoda

Family: Palinuridae

Common name: Spiny lobster

Distribution: Eastern Pacific Ocean

Habitat: Rocky substrates, at depths of up to 65 m

90

Description: This species has two large, spiny antennae, but no large claws on its legs. It may grow up to 60 cm long, but does not usually exceed 30 cm. Males can weigh up to 12 kg. The upper side of the animal is brownish red. The legs are a similar color, but with one or more lighter streaks running along their length

Nutritional Facts

Proximate Composition (g/100g)

Protein	CHO	Fiber	Fat
26.0	3.1	0	1.9

Fatty Acids and Cholesterol (g/100g)

SFA	MUFA	PUFA	Cholesterol*
0.3	0.4	0.8	90

* mg

Minerals (mg/100 g)

Na	K
227	208

Homarus gammarus Linnaeus, 1758

Order: Decapoda

Family: Nephropidae

Common name: European lobster, common lobster

Distribution: Northeastern Atlantic Ocean; Mediterranean Sea; Norwegian fjords Tysfjorden and Nordfolda

Habitat: Rocky substrata, living in holes and excavated tunnels from the lower shore to about 60 m depth

Description: A large lobster species that can grow up to 1 m in length, but 50 cm is more common. It is blue-colored above with coalescing spots and yellowish coloring below. The first pair of walking legs carry massive (but slightly unequal) pincers, which can be formidable and dangerous. The body lacks strong spines or ridges and is only slightly granular.

91

Homarus americanus H. Milne-Edwards, 1837

Order: Decapoda

Family: Nephropidae

Common name: American lobster

Distribution: Atlantic coast of North America

Habitat: Cold, shallow waters with rocks; depth range of 4–480 m

Description: This species commonly reaches 200–610 mm long and weighs 0.45–4.08 kg in weight, but it has been known to weigh as much as 20 kg, making this the heaviest and longest decapod crustacean in the world.

Nutritional Facts

	Homarus gammarus	*H. americanus*
Moisture (%)	78.1	79.2
Ash (%)	1.8	1.8
Protein (%)	18.3	17.1
Fat (%)	0.3	0.7
Cholesterol (mg/100 g WW)	36.6	43.2
Energy (Kcal/100 g WW)	87.0	82.3
SFA (%)	22.7	21.6
MUFA (%)	28.8	31.5
PUFA (%)	43.6	42.0

WW: wet weight; SFA: saturated fatty acids; MUFA: monounsaturated fatty acids; PUFA: polyunsaturated fatty acids.

Source: Data from Barrento et al. (2009).

Linuparus somniosus Berry & George, 1972

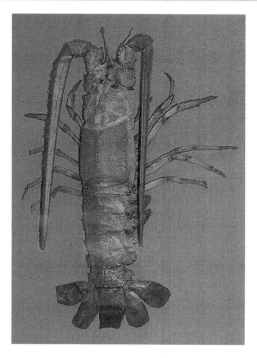

Order: Decapoda

Family: Palinuridae

Common name: African spear lobster

Distribution: Off the east coast of Africa from Kenya to Natal, South Africa

Habitat: Depth ranges from 216 to 375 m; on rough substrate with sand and mud

Description: Submarginal posterior groove of carapace of this species is much wider medially than laterally. Vestigial pleopods are present on first abdominal segment of female. Maximum total body length is 35 cm, and carapace length is 14 cm.

Nutritional Facts

Proximate Composition (%)

Ash	Water	Protein	Fat
1.05	83.45	12.29	0.51

Source: Data from Suseno et al. (2013).

6
MOLLUSKS

OYSTERS

Crassostrea gryphoides (Schlotheim 1813)

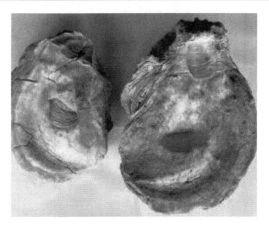

Phylum: Mollusca

Class: Bivalvia

Order: Ostreoida

Family: Ostreidae

Common name: Estuarine oyster or giant oyster

Distribution: India, Pakistan

Habitat: Zero-tide level on creeks, estuaries, backwaters, and muddy low tidal areas

Description: Shells are elongated. Anterior region is narrow, and posterior region is broad. While lower valve is slightly concave, heavy, and thick, upper valve is flat and less thick. Hinge is long, deeply grooved, and curved to the right or left. Adductor muscle scar is white, roundish or bean-shaped, and dorsally placed. Life span of this species is up to 40 years.

Crassostrea rivularis (Gould 1861)

Common name: Jinjiang oyster or Chinese oyster

Distribution: Throughout the western Pacific (native to Japan, China, India, and Pakistan)

Habitat: Intertidal hard grounds and muddy creeks of warm estuaries

Description: Shells are elongated. Lower valve is deep and cup-shaped. Upper valve is opercular and flat. Muscle scar is displaced both toward dorsolateral edge and toward tip. A promyal chamber is present on the right side.

Crassostrea madrasensis (Preston)

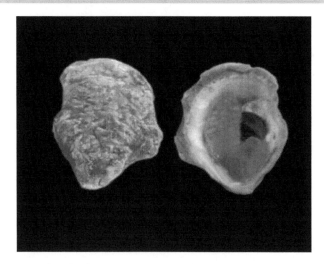

Common name: Indian backwater oyster

Distribution: India, Sri Lanka

Habitat: Estuaries, backwaters, mangroves

Description: Maximum length, width, and height of the shell are 65 mm, 49 mm, and 35 mm, respectively. Although the maximum weight of the species is 37 g, fresh meat weight is only 3 g.

Crassostrea gigas (Thunberg, 1793)

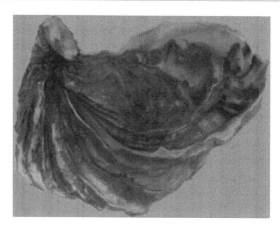

Common name: Pacific oyster, Japanese oyster, or Miyagi oyster

Distribution: Native to the Pacific coast of Asia and introduced species in North America, Australia, Europe, and New Zealand

Habitat: Intertidal and subtidal zones; hard or rocky surfaces in shallow or sheltered waters up to 40 m deep; muddy or sandy areas

Description: Shell of this species varies widely with the environment where it is attached. Its large, rounded, radial folds are often extremely rough and sharp. Two valves of the shell are slightly different in size and shape, with the right valve being moderately concave. Shell color is variable, usually pale white or off-white. Mature specimens vary from 80 mm to 400 mm long.

Saccostrea cucullata (Born, 1778)

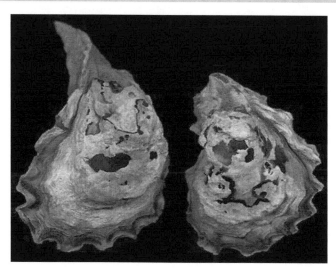

Common name: Hooded oyster

Distribution: Indian Ocean and tropical west Pacific Ocean. In East Africa, its range includes Somalia, Kenya, Tanzania, and the Seychelles

Habitat: Intertidal zone, and at depths down to about 15 m

Description: Shape of the shell is circular or oblong or roughly oval, often with an irregular outline. Maximum size is 12 cm. Valves are thick and solid. Lower valve is convex and has no sculpturing near the umbo, which is fixed to the substrate. Upper valve is flat and smaller than the lower valve. It may have wide, sometimes spiny ribs. Margins of the valves are pleated and fit together neatly. A single large adductor muscle holds the valves together. Color is purplish-brown on the outside of the valves; inside is white rimmed with black.

Nutritional Facts (Oysters) (per 100 g of Raw Edible Portion)

Nutrients	Amount
Macronutrients	
Protein	10.8 g
Lipid	1.3 g
Carbohydrate	2.7 g
Vitamins	
Vitamin A	75 µg
Vitamin D	1 µg
Vitamin F	0.85 mg
Vitamin B	0.15 mg
Vitamin B_2	0.19 mg
Vitamin B_6	0.16 mg
Vitamin B_{12}	17 µg
Minerals	
Sodium	510 mg
Potassium	260 mg
Calcium	140 mg
Iron	5.7 mg
Zinc	59.2 mg
Iodine	60 µg
Selenium	23 µg

Source: Data from Holland et al. (1993).

MUSSELS

Mytilus edulis Linnaeus 1758

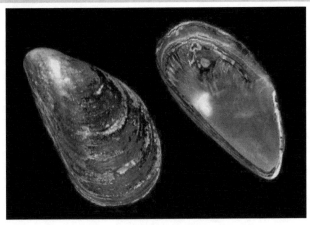

98

Order: Mytiloida

Family: Mytilidae

Common name: Blue mussel

Distribution: Circumpolar in boreal and temperate waters, in both the southern and northern hemispheres extending from the Arctic to the Mediterranean in the northeast Atlantic

Habitat: Intertidal areas attached to rocks and other hard substrates

Description: Shape of the shell is smooth, triangular, and elongated with rounded edges. Fine concentric growth lines are present. Color of the shell is purple, blue, or sometimes brown, occasionally with radial stripes. Outer surface of the shell is covered by the periostracum, which, as it erodes, exposes the colored prismatic calcitic layer. Blue mussels are semisessile, having the ability to detach and reattach to surface.

Mytilus galloprovincialis Lamarck, 1819

Common name: Mediterranean mussel

Distribution: Europe (Mediterranean Sea and the Black Sea); Atlantic coast (in Portugal, north to France and the British Isles); Northern Pacific (along the coast of California); Asian coast throughout Japan, including Ryukyu Islands, as well as in North Korea and around Vladivostok in Russia

Habitat: Subtidal zones

Description: Shell, which has a maximum length of 140 mm, is smooth, brittle, oval, subtriangular, or pear shaped. Umbones, which are beaked and down-turned, are prominent, pointed, and slightly curved ventrally. Fine concentric lines are present. Color of the shell is blue to deep purplish-black. Periastracum is light brown to blue-black. This species is a filter feeder.

Modiolus (Modiolus) modiolus (Linnaeus, 1758)

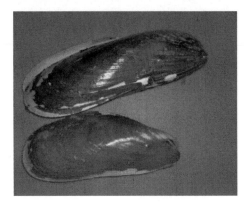

Common name: Horse mussel, bearded mussel, fan mussel, giant horse mussel

Habitat: Low-tide mark to depths of 50 m; hard substrates, including shells and stones and the byssus threads of other mussels

Distribution: Atlantic and Pacific oceans, from the Arctic to subtropics; subtidal and intertidal (mostly on the underside of rocks); muddy bottoms with shell hash

Description: This mussel is longer than high; umbones are not quite terminal. It attaches by byssal threads to firm substrates, its length is about 2–3 times the height, and the periostracum is usually brown (or black in large individuals) and grown out into long, soft bristles, especially in small individuals and near the margin of the shell. Up to 15 cm long.

Lithophaga lithophaga (Linnaeus, 1758)

Common name: Date mussel or date shell

Distribution: Northeast Atlantic Ocean, the Mediterranean Sea, and the Red Sea; Adriatic coast of Croatia and Montenegro

Habitat: Bores into marine rocks

100

Description: Shells of this species are long (5 cm) and narrow with parallel sides. Oldest individuals are found at depths of 1–5 m. Maximum growth (75%) has been observed from the end of spring to the beginning of summer. Increase in length of the borehole, which is 1.5 times greater than the length of *L. lithophaga*, is continuous and occurs at a faster rate in winter.

Nutritional Facts (Mussels) (per 100 g of Raw Edible Portion)

Nutrients	Amount
Macronutrients	
Protein	12.1 g
Lipid	1.8 g
Carbohydrate	2.5 g

Vitamins	
Vitamin A	NA
Vitamin D	TR
Vitamin F	0.74 mg
Vitamin B	0.02 mg
Vitamin B$_2$	0.35 mg
Vitamin B$_6$	0.08 mg
Vitamin B$_{12}$	19 µg
Minerals	
Sodium	290 mg
Potassium	320 mg
Calcium	38 mg
Iron	5.8 mg
Zinc	2.5 mg
Iodine	140 µg
Selenium	51 µg

NA: data not available; TR: traces.
Source: Data from Holland et al. (1993).

CLAMS

Ensis directus Conrad, 1843

Order: Veneroida

Family: Pharidae

Common name: Razor clam, American jackknife clam

Distribution: North American Atlantic coast, Europe

Habitat: Sand and mud of intertidal or subtidal zones in bays and estuaries

Description: It has a thin, streamlined shell that is slightly curved. Color of the shell ranges from yellowish to dark brown. Length of this species is about six times its width. It can grow up to 25 cm. It has a coating around its shell to protect

101

it from eroding in the mud or sand. Two valves of the shells that are identical are connected by an elastic ligament. Strong foot helps in burrowing and swimming.

Arctica islandica Linnaeus, 1767

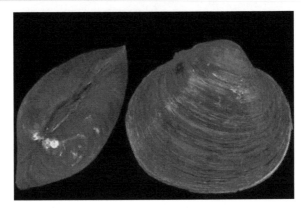

Order: Veneroida

Family: Arcticidae

Common name: Ocean quahog, black clam

Distribution: Native to the North Atlantic Ocean; Britain and Ireland

Habitat: Just below the low-water level to depths of about 500 m; buried in sand and muddy sand in subtidal zones

Description: It is a typical cockle-shaped bivalve, and the two halves of its hinged, rounded shell are thick, glossy, and dark brown in color. It is a long-lived animal and is quite large, growing up to 13 cm across. Pallial line has no indentation or sinus. Its siphon keeps water flowing across the animal, so that it can breathe, capture food, and expel waste.

Donax trunculus Linnaeus, 1758

Order: Veneroida

Family: Donacidae

Common name: Tellina or abrupt wedge shell

Distribution: Coasts of Western Europe and Northwestern Africa

Habitat: Near the shore as well as between 0 and 2 m below the surface

Description: Shell of this species is wedge shaped, equivalve, no longer than 150 mm, with the posterior longer than the anterior. It is usually muted when compared to the polychromic shell interior. Interior is smooth, and the pallial line is strong. This species has two cardinal teeth in each valve and a heterodont hinge plate. Anterior and posterior pedal retractor muscles, as well as pedal protractors and elevators, are all present. Foot is large, muscular, and bladelike.

Meretrix meretrix (Linnaeus, 1758)

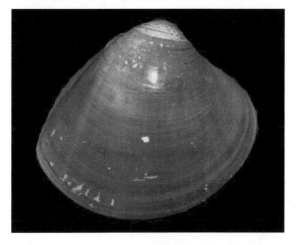

Order: Veneroida

Family: Veneridae

Common name: Asiatic hard clam

Distribution: South and Southeast Asia

Habitat: Estuarine and coastal ecosystems

Description: Shell, which has a total length of 75 mm, is trigonal ovate. Posterior margin is slightly more pointed than the anterior. Anterior dorsal margin is nearly straight, and posterior dorsal margin is slightly convex. Valves are equal in size and shape. Sculpture consists of growth lines only. Exterior color is white. There may be light purple or pink pigmentation along co-marginal and/or radial lines. Periostracum is dirty gray. Interior is white. On the right valve, there are three cardinal teeth. The posterior adductor muscle scar is rounder than the anterior, and the pallial sinus is shallow.

Meretrix lusoria Roeding, 1798

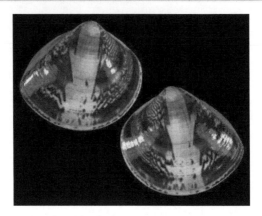

Common name: Hard clam, common Orient clam

Distribution: Japan and the southern coast of Korea

Habitat: Marine

Description: Overall shape of the shell is trigonal ovate. Posterior margin is more pointed than the anterior, and ventral margin is rounded. Valves are the same size and shape (equivalve). Sculpture consists of growth lines only. Umbo is slightly anterior. Exterior surface is white and typically has brown tent marks radial stripes, and/or co-marginal stripes. Periostracum is dark brown or dirty gray. Interior color is white. On the left and right valves, there are three cardinal teeth. Posterior adductor muscle scar is slightly more rounded than the anterior.

Venerupis philippinarum (A. Adams & Reeve, 1850)

Common name: Manila clam, Japanese littleneck clam

Distribution: Native from southern Siberia to China; and from the British Columbia coast to the central coast of California and Marina del Rey (Los Angeles)

Habitat: Mid- to low intertidal zone in bays and estuaries

104

Description: Shell is elongate, oval, and sculptured with radiating ribs with a maximum length of about 6 cm. Color is grayish, greenish, brownish, yellow, or buff with distinct dark or light coloring and triangular mottled markings. Foot of live specimens is orange. Siphons are short and are mostly fused.

Saxidomus giganteus (Deshayes, 1839)

Common name: Butter clam, Washington clam, money shell

Distribution: Aleutian Islands and southeastern Bering Sea, Alaska, to San Francisco Bay, California

Habitat: Sheltered sand, sandy mud, and gravel beaches; low intertidal to 40 m

Description: This species, which measures 15 cm, has a large, black external hinge ligament and well-developed concentric ridges. Shell is only slightly longer than it is high. It has a true hinge plate with three cardinal teeth in each valve. The valves have a smooth but not glossy interior with a pallial sinus and a continuous pallial line. Outside of the shell may be white, or may be stained by iron sulfide in anoxic sediments. Siphons are moderately long (4 cm).

Protothaca staminea (Conrad, 1837)

Common name: Pacific littleneck clam, ribbed carpet shell

Distribution: Tropical West America

Habitat: Lower half of the intertidal down to 10 m depth; in stable sand, packed mud, or gravel-clay mixtures; in gravelly sediments among rocks on the open coast

Description: This species, which has a total length of 7 cm, has a porcelain-type (chalky) shell. Umbo is anterior to the midline. Valves are more or less oval or heart shaped, and they are similar in size. This species has many radial ribs. Siphons are fused all the way to the end and have black tips. Valves are whitish inside without a purple stain and have a row of very small teeth along the inside of the ventral margin. Shell is whitish, gray, yellowish, or brown with little periostracum on the outside, and they may have zigzag brown markings or brown splotches.

Venerupis decussata (Linnaeus, 1758)

Common name: Cross-cut carpet shell, checkered carpet shell

Distribution: From the North Sea and British Isles southward to the Mediterranean Sea and North Africa

Habitat: Buried in soft substrates, sand, muddy sand, gravel, or clay, on the lower shore and at depths down to a few meters in the intertidal zone

Description: Shell of this species is broadly oval or square in shape and is cream, yellowish, or light brown in color, often with darker markings. Sculpture of the shell consists of concentric grooves and bold radiating ridges. It grows up to 7.5 cm in length. Each valve has three teeth, one on the left and two on the right, which are bifid. Inside of the shell is polished white with an orange tint, occasionally with purple over a wide area below the umbones.

Mercenaria mercenaria (Linnaeus, 1758)

Common name: Hard clam

Distribution: East coast of the United States

Habitat: Sandy bottoms of intertidal zones

Description: It has a large, heavy shell that ranges from being a pale brownish color to shades of gray and white. Exterior of the shell, except nearest the umbo, is covered with a series of growth rings. Interior of the shell is colored a deep purple around the posterior edge and hinge.

Mactra sachalinensis Schrenck, 1862

Order: Veneroida

Family: Mactridae

Common name: Hen clam or surf clam

Distribution: East coast of the United States, northern Japan

Habitat: Buried in coarse or fine sand; offshore as well as in the low intertidal and surf zones

Description: It grows to a maximum length of 20 cm. Its shell is generally roughly trigonal in shape, with concentric striated ridges and a yellowish-brown periostracum covering. It has a partially exposed spisuloid ligament.

Mya arenaria Linnaeus, 1758

Order: Myoida

Family: Myidae

Common name: Soft-shell clam or long-necked clam

Distribution: Native to eastern North America; Iceland, Norway, Sweden, Baltic Sea, Denmark, Faeroe Islands, Ireland, England, Atlantic France, Spain, Mediterranean France, western Sicily, northern Adriatic Sea, and the Black Sea

Habitat: Buried up to 30 cm below the surface in sand, mud, and clays, often in mixtures with coarse gravel; upper intertidal zone in bays and estuaries; low intertidal and shallow subtidal zones and deeper water

Description: This species has an oval shell that is rounded at the head end and slightly pointed at the hind end. It can reach up to 15 cm in length. Shells are blue-white or brownish-white. Shell surface is roughly sculpted along an uneven, concentric growth line. The two halves of shell are conspicuously different.

Panopea generosa Gould, 1850

Order: Myoida

Family: Hiatellidae

Common name: Geoduck clam

Distribution: Native to the west coast of North America; worldwide nontropical

Habitat: Coastal wetlands

Description: This is the largest burrowing clam in the world, weighing an average of 0.5–1.5 kg. But specimens weighing over 7.5 kg and as much as 2 m in length have also been reported. Shell of this clam ranges from 15 cm to over 20 cm in length, but the extremely long siphons make the clam itself much longer than this. "Neck" or siphons alone may be 1 m in length.

Nutritional Facts (Clams) (for a Serving Size of 227 g)

Nutrients	Amount
Cholesterol	77.2 mg
Protein	29 g
Fats	
Total fat	2.2 g
Saturated fat	0.2 g
Monounsaturated fat	0.2 g
Polyunsaturated fat	0.6 g
Fatty Acids	
Omega-3 fatty acids	449 mg
Omega-6 fatty acids	36.3 mg

Carbohydrates	
Total carbohydrates	5.5 g
Vitamins	
Vitamin A IU	681 IU
Vitamin A (retinol activity equivalent)	204 µg
Vitamin E (alpha tocopherol)	0.7 mg
Riboflavin	0.5 mg
Niacin	4 mg
Thiamin	0.2 mg
Vitamin B_6	0.1 mg
Folate	36.3 µg
Vitamin K	0.5 µg
Dietary folate equivalents	36.3 µg
Vitamin C	29.5 µg
Pantothenic acid	0.8 mg
Choline	148 mg
Vitamin D	9.1 IU
Vitamin B_{12}	112 µg
Minerals	
Calcium	104 mg
Iron	31.7 mg
Magnesium	20.4 mg
Phosphorus	384 mg
Potassium	713 mg
Zinc	3.1 mg
Copper	0.8 mg
Selenium	55.2 µg
Sodium	127 mg

Source: Data from Extension: America's Research-Based Learning Network (2013).

109

SCALLOPS

Chlamys farreri (Jones & Preston, 1904)

Phylum: Mollusca

Class: Bivalvia

Order: Ostreoida

Family: Pectinidae

Common name: Japanese scallop or Chinese scallop

Distribution: Pacific Asian subtropical

Habitat: Gravel and pebbled grounds

Description: Both valves of this species are convex; usually, the left door is a bit more curved. Front ear is greater than the posterior ear. Outside is covered with numerous spiny ribs or beads, wherein the beads are again occupied by finer radial ribs. Inner edge has fine radial ribs. Shells are conspicuously colored (orange, yellow, pink, purple, brown, or white). Rows of eyes are present as black pigment spots in the mantle fold in living animals. Shell has a maximum height of 112 cm.

Patinopecten yessoensis Jay, 1857

Common name: Yesso scallop, giant Ezo scallop, Ezo giant scallop

Distribution: Eastern Asian coast (China, Korea, Japan, and Sakhalin); Kamchatka Peninsula and Aleutian Islands

Habitat: Sheltered, shallow bays and inlets adjacent to rocky shores, out to a depth of 30–40 m in more open sea areas

Description: Shells are large (10–22 cm long) and almost circular. Umbones are in center, between two almost equal ears. Exterior of right valve is whitish, with 20 broad, flattened ribs. Exterior of left valve is purplish brown with 20 coarse radial ribs. Interior valve is whitish and furrowed, with a single adductor muscle scar.

Pecten albicans (Schröter, 1802)

Common name: Japanese baking scallop

Distribution: Japanese and the South China Seas

Habitat: Shallow inshore reef areas, at depths of 40–115 m

Description: This species has a shell reaching a size of 95 mm, with about 12 radiating ribs. Color of the surface usually ranges from light brown to dark brown, but it may also be orange or purple. Lower valve of this species is less convex, and upper valve is flat.

Nutritional Facts (Scallops) (for a Serving Size of 113.40 g)

Nutrients	Amount
Protein	23.11 g
Carbohydrates	
Dietary fiber	0.00 g
Starch	0.00 g

Fats	
Total fat	4.48 g
Saturated fat	0.77 g
Mono fat	1.72 g
Poly fat	1.31 g
Vitamins	
Vitamin A IU	211.03 IU
Vitamin A retinol activity equivalent (RAE)	45.82
Niacin: B_3	1.50 mg
Vitamin C	3.93 mg
Vitamin E IU	2.81 IU
Folate	20.94 µg
Minerals	
Calcium	34.28 mg
Copper	0.07 mg
Iron	0.40 mg
Magnesium	77.12 mg
Manganese	0.10 mg
Phosphorus	302.05 mg
Potassium	444.46 mg

Continued

111

Nutrients	Amount
Sodium	261.63 mg
Zinc	1.31 mg
SFA	
Fatty Acids	
16:0 Palmitic	0.46 g
Mono Fats	
18:1 Oleic	1.68 g
Poly Fats	
18:2 Linoleic	0.88 g
18:3 Linolenic	0.07 g
Omega-3 fatty acids	0.35 g
Omega-6 fatty acids	0.07 g
Amino Acids	
Alanine	1.39 g
Arginine	1.68 g

Aspartate	2.22 g
Cystine	3.03 g
Glutamate	3.14 g
Glycine	1.44 g
Histidine	0.44 g
Isoleucine	1.00 g
Leucine	1.62 g
Lysine	1.72 g
Methionine	0.52 g
Phenylalanine	0.83 g
Proline	0.94 g
Serine	1.03 g
Threonine	0.99 g
Tryptophan	0.26 g
Tyrosine	0.74 g
Valine	1.01 g

Source: Data from SELFNutrition Data (2014).

CONCHES

Strombus gigas Linnaeus, 1758

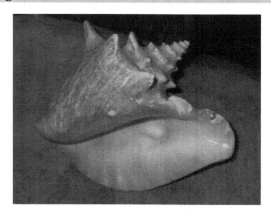

Phylum: Mollusca

Class: Gastropoda

Order: Mesogastropoda

Family: Strombidae

Common name: Queen conch

Distribution: Caribbean Sea, Gulf of Mexico, Bermuda, Brazil

Habitat: Sand, seagrass bed, and coral reef habitats; shallow water

Description: Adult animal has a very large, solid, and heavy spiral-shaped shell, with knoblike spines on the shoulder, a flared thick, outer lip, and a characteristic pink-colored aperture (opening). It has a long snout, two eyestalks with well-developed eyes, additional sensory tentacles, a strong foot, and a corneous, sickle-shaped operculum. This species grows to a maximum length of 30 cm and weight of 2 kg.

Babylonia spirata (Linnaeus, C., 1758)

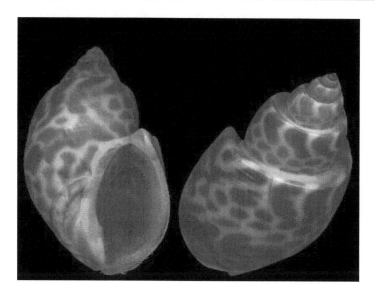

Order: Neogastropoda

Family: Babyloniidae

Common name: Spiral Babylon

Distribution: Tropical Pacific in the Indo-Pacific region stretching from the Eastern Indian Ocean east to the northern shores of Australia and New Zealand

Habitat: Shallow sandy or muddy seawater

Description: This species has a thick and heavy shell with a size range of 33–75 mm. It has an almost flush-sided body whorl (outside shell around its imaginary axis). A spire seems pushed down into the body, separated by a deep channel wrapping around the body whorl. The shell has rings of alternating brown and cream colors spiraling around the body.

Nutritional Facts

Proximate Composition (%)

Protein	Carbohydrate	Lipid	Amino Acids
53.86	16.85	9.30	9.91*

* *Micrograms per gram.*

Amino Acids (mg/g)

Isoleucine	3.081
Valine	2.066
Lysine	1.012
Phenylalanine	1.011
Leucine	0.889
Methionine	0.766
Proline	0.665
Tryptophan	0.221
Glutamic acid	0.112
Alanine	0.088
Total	9.911

Source: Data from Periyasamy et al. (2011).

113

Bullacta exarata (Philippi, 1849)

Order: Cephalaspidea

Family: Haminoeidae

Common name: Korean mud snail

Distribution: Native to coastlines of the South and East China Seas from Hainan to the Bohai Sea in northeastern China; western coast and south coast of Korea and Japan

Habitat: Intertidal, including supratidal and subtidal zones

Description: Shell is bullate, fairly thick, white, and spirally striate, with a well-developed periostracum. Columella is smooth and simple. Aperture extends for the whole length of the shell, and it is narrower above than below. Apertural lip extends upward beyond the apex of the shell. The height of the shell is 8 mm, and the width of the shell is 6 mm. Foot and the epipodia are (in alcohol) of a pale green color. Foot is short and truncate before and behind. Eyes are minute and quite invisible on the surface.

Nutritional Facts (Conch) (per 100 g)

Total fat	1.5 g
Saturated fat	0.5 g
Monounsaturated fat	0.4 g
Polyunsaturated fat	0.3 g
Cholesterol	82.6 mg
Sodium	194.3 mg
Potassium	207 mg
Carbohydrates	2.2 g
Protein	33.4 g

Vitamins and Minerals

Vitamin A	8.9 IU
Vitamin B$_6$	0.1 mg
Vitamin B$_{12}$	6.7 µg
Vitamin E	8 mg
Vitamin K	0.3 µg
Calcium	124.5 mg
Iron	1.8 mg
Magnesium	302.3 mg
Phosphorus	275.6 mg
Zinc	2.2 mg
Copper	0.6 mg
Selenium	51.2 µg
Retinol	8.9 µg
Thiamine	0.1 mg
Riboflavin	0.1 mg
Niacin	1.3 mg
Folate	227.3 µg
Choline	102.9 mg

Source: Data from Daily Diet Guide (2006).

114

ABALONES

Haliotis spadicea Donovan, 1808

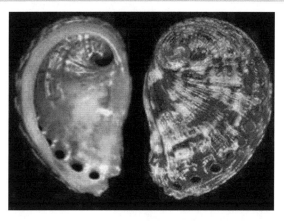

Order: Archaeogastropoda

Family: Haliotidae

Common name: Blood-spotted abalone

Distribution: Indian Ocean off South Africa

Habitat: Rocky shores

Description: Exterior of the shell of this species is reddish-purple in color, often with some white blotches. Shell has between five and eight open respiratory pores along the margin. These holes collectively make up what is known as the selenizone, which forms as the shell grows. Shell grows to 70 mm in length.

Haliotis cracherodii Leach, 1814

Common name: Black abalone

Distribution: Along the Pacific coast from Mendocino County, California, USA, to Cabo San Lucas, Baja California, Mexico

115

Habitat: Crevices, cracks, and holes of intertidal and shallow subtidal rocks

Description: Shell is smooth black or slate blue, and inside is pearly white. Mantle and foot are black. There are 5–9 open respiratory pores along the left side of the shell and spiral growth lines on the rear. Tentacles are present surrounding the foot and extending out of the shell. Maximum length and weight are 20 cm and 800 g, respectively.

Haliotis tuberculata Linnaeus, 1758

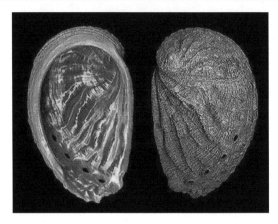

Common name: European edible abalone, green ormer snail

Distribution: European waters from the Mediterranean Sea to as far north as the English Channel; in the Atlantic Ocean off the Canary Islands and West Africa

Habitat: Rocky shores

Description: Shell of this species grows as large as 10 cm in length and 6.5 cm in width. This flattened, oval shell is an ear-shaped spiral with a mottled outer surface. The inner surface of the shell has a thick layer of iridescent pearl. Its large and muscular foot has numerous tentacles at the epipodium (the lateral grooves between the foot and the mantle).

Haliotis rufescens Swainson, 1822

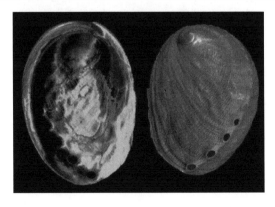

Common name: California red abalone

Distribution: Along the west coast of North America, from Oregon, United States, to Baja California, Mexico

Habitat: Rocky areas with kelp; intertidal zone to water of 30 m depth

Description: Its shell length can reach a maximum of 31 cm, making it the largest species of abalone in the world. Shell is large, thick, and dome shaped. It is usually a brick red color externally. Shell has three or four oval holes or respiratory pores. Inside of the shell is strongly iridescent and has a large central muscle.

Haliotis rubra W.E. Leach, 1814

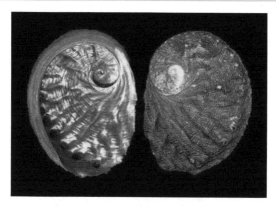

Common name: Black lip abalone

Distribution: Endemic to Australia; Tasmania

Habitat: In crevices, caves, and fissures and on vertical rock surfaces, from low tide to 40 m

Description: Specimens are sculptured with irregular, coarse radial or oblique folds, and fine scaly spiral ribs. Shell angled at the row of siphonal holes, the last 6–8 of which are open. Wide groove is seen below angle. External color is terracotta red occasionally streaked with light green, and interior is silvery iridescent. Size of the shell varies between 35 mm and 200 mm. Edge of the foot is black.

Haliotis gigantea Gmelin, 1791

Common name: Siebold's abalone

Distribution: Endemic to the waters off Japan and Korea

Habitat: Marine

Description: The species is large, rounded, and heavier, and the color is mainly brick red, with paler radial bands.

A typical feature of this species is, at the extremity of the row of holes, the triangular shape of the margin. Row of holes runs along an angled crest. Dorsal side is like an old folded cliff. Few holes are open, usually less than six. Maximum size of the shell is 220 mm.

Haliotis discus hannai Ino, 1953

Common name: Pacific abalone, Japanese disc abalone, Japanese abalone

Distribution: East and west coasts of Korea; Liaodong and Shandong peninsula in China; Japan

Habitat: Beds rich in algal vegetation of kelp and crustose coralline algae

Description: Shell of this species has a low open spiral structure, and it is characterized by several open respiratory pores in a row near the shell's outer edge. Thick inner layer of the shell is composed of nacre (mother-of-pearl), which may be highly iridescent, giving rise to a range of strong changeable colors.

Nutritional Facts (Abalones)
(for a Serving Size of 85 g)

Nutrients	Amount
Total Fat	0.6 g
Saturated fat	0.1 g
Monounsaturated fat	0.1 g
Polyunsaturated fat	0.1 g
Cholesterol	72.3 mg
Sodium	255.9 mg
Potassium	212.5 mg
Carbohydrates	5.1 g
Protein	14.5 g
Vitamins and Minerals	
Vitamin A	1.7 IU
Vitamin B_6	0.1 mg
Vitamin B_{12}	0.6 µg

Continued

Nutrients	Amount
Vitamin C	1.7 mg
Vitamin E	3.4 mg
Vitamin K	19.6 µg
Calcium	26.4 mg
Iron	2.7 mg
Magnesium	40.8 mg
Phosphorus	161.5 mg
Zinc	0.7 mg
Copper	0.2 mg
Selenium	38.1 µg
Retinol	1.7 µg
Thiamine	0.2 mg
Riboflavin	0.1 mg
Niacin	1.3 mg
Folate	4.3 µg
Choline	55.3 mg

Amino Acids	
Tryptophan	0.2 g
Threonine	0.6 g
Isoleucine	0.6 g
Leucine	1 g
Lysine	1.1 g
Methionine	0.3 g
Cystine	0.2 g
Phenylalanine	0.5 g
Tyrosine	0.5 g
Valine	0.6 g
Arginine	1.1 g
Histidine	0.3 g
Alanine	0.9 g
Aspartic acid	1.4 g
Glutamic acid	2 g
Glycine	0.9 g
Proline	0.6 g
Serine	0.7 g

Source: Data from Eatthismuch.com (2013).

LIMPET

Cellana exarata (Reeve, 1854)

Phylum: Mollusca
Class: Gastropoda
Order: Archaeogastropoda

Family: Nacellidae
Common name: Blackfoot opihi, blackfoot limpet

Distribution: Endemic to the islands of Hawaii

Habitat: From the spray zone to the coralline algae belt of the mid-intertidal zone

Description: Animal grows to a maximum size of only 5 cm. Both the shell and the foot of the animal are black in color. Interior is dark-gray. Close-set ribs with narrower ones between, not extending far beyond shell margin.

Nutritional Facts: Limpet (for a Serving Size of 53 g)

Nutrients	Amount
Macronutrients	
Fat	0
Cholesterol	117 mg
Carbohydrates	1 mg
Protein	9 mg
Mineral	
Sodium	350 mg

Source: Data from Dietfacts.com.

SQUIDS

Todarodes pacificus (Steenstrup, 1880)

Phylum: Mollusca

Class: Cephalopoda

Order: Teuthida

Family: Ommastrephidae

Common name: Japanese flying squid, Japanese common squid, Pacific flying squid

Distribution: Northern Pacific Ocean, Japan, China up to Russia, Bering Strait, Alaska, Canada, Vietnam

Habitat: Upper layers of the ocean

Description: Mantle encloses the visceral mass. It has eight arms and two tentacles with suction cups along the backs. In between the arms sits the mouth, or beak. Inside the mouth is a tooth–tongue-like appendage called the radula. Its ink sac is used as a defense mechanism against possible predators. This species may weigh up to 0.5 kg. Mantle length in females can be up to 50 cm. Males are smaller.

Ommastrephes bartramii (Lesueur, 1821)

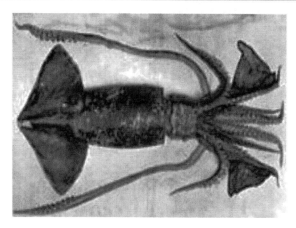

Common name: Red squid, red ocean squid, neon flying squid

Distribution: Subtropical and temperate waters of the Pacific, Atlantic, and Indian Oceans

Habitat: Beneath the surface near cold waterfronts, at depths of 0–700 m

Description: This species has an elongated silver-colored band in the middle of the ventral side of the mantle.

Adult male may reach a maximum length of 45 cm. Adult female is much larger, usually having a mantle length of around 50 cm with the maximum known length being 60 cm. Arms have 9–27 suckers on the ventral sucker series and 10–25 suckers on the dorsal sucker series. The hectocotylus develops from the left or right fourth arm. It also has 4–7 toothed suckers on the tentacular club.

Illex argentines (Castellanos, 1960)

Common name: Argentine short fin squid

Distribution: Western South Atlantic

Habitat: Oceanic and neritic species occurring from the surface to about 800 m depth

Description: Mantle is long, muscular, and widest at midpoint. Fins are muscular and relatively short and broad. All arms of males are significantly longer than in females. Right (or left) of fourth arm pair is hectocotylized with suckers, and stalks are modified into suckerless knobs.

Uroteuthis chinensis (Gray, 1849)

Order: Teuthida

Family: Loliginidae

Common name: Mitre squid

Distribution: Indo-West Pacific

Habitat: Coastal water to depths of 170 m

Description: Proximal margin of sucker rings of arms of this species has a semi-crescent plate. Two rows of papillae are present in hectocotylus. Tentacular clubs of tentacles are expanded, and suckers are in four series. Mantle is elongated and pointed posteriorly. Fins in adults are rhomboidal and longer than broad, tapering posteriorly. Maximum length is 30 cm.

Nutritional Facts (Squid) (for a Serving Size of 85 g)

Nutrients	Amount
Total fat	1.2 g
Saturated fat	0.3 g
Monounsaturated fat	0.1 g
Polyunsaturated fat	0.4 g
Cholesterol	198 mg
Sodium	37.4 mg
Potassium	209.1 mg
Carbohydrates	2.6 g
Protein	13.2 g
Vitamin A	8.5 IU
Vitamin B$_{12}$	1.1 µg

Vitamin C	4 mg
Vitamin E	1 mg
Calcium	27.2 mg
Iron	0.6 mg
Magnesium	28.1 mg
Phosphorus	187.9 mg
Zinc	1.3 mg
Copper	1.6 mg
Manganese	0 mg
Selenium	38.1 µg
Retinol	8.5 µg
Riboflavin	0.4 mg
Niacin	1.8 mg
Folate	4.3 µg
Choline	55.3 mg
Tryptophan	0.1 g
Threonine	0.6 g
Isoleucine	0.6 g
Leucine	0.9 g
Lysine	1 g
Methionine	0.3 g
Cystine	0.2 g
Phenylalanine	0.5 g
Tyrosine	0.4 g
Valine	0.6 g
Arginine	1 g
Histidine	0.3 g
Alanine	0.8 g
Aspartic acid	1.3 g
Glutamic acid	1.8 g
Glycine	0.8 g
Proline	0.5 g
Serine	0.6 g

Source: Data from EatThisMuch.com (2014).

CUTTLEFISH

Sepia esculenta Hoyle, 1885

Order: Sepioloida

Family: Sepiidae

Common name: Golden cuttlefish

Distribution: Northwest and western central Pacific

Habitat: Demersal; sandy substrates; sea grass beds

Description: A distinct golden line is seen around the base of the fins. It also has numerous wavy lines across the body. It grows to a maximum length of 18 cm. It is able to shoot a cloud of black ink at predators when threatened.

Sepiella japonica Sasaki, 1929

Common name: Japanese spineless cuttlefish

Distribution: Pacific northeast

Habitat: Benthic, continental shelf

Description: This species attains a maximum mantle length of 200 mm and body weight of 800 g. Body of this species is dorsoventrally flattened. Mantle, which has a length of 20 cm, is elongated oval and much larger than the head. Anterior dorsal edge of the mantle has a tongue-like projection. A row of several large, brightly colored spots extends along the fin base. Fins are narrow, extending from the anterior to posterior edge of the mantle. Arms have 2–4 rows of suckers. Fourth arm is considerably broader than the others, with a muscular swimming membrane. Tentacles are retractile. Club suckers are in 20 rows.

Nutritional Facts (Cuttlefish) (for a Serving Size of 85 g)

Nutrients	Amount
Total fat	1.2 g
Saturated fat	0.2 g
Monounsaturated fat	0.1 g
Polyunsaturated fat	0.2 g
Cholesterol	190.4 mg
Sodium	632.4 mg
Potassium	541.4 mg
Carbohydrates	1.4 g
Protein	27.6 g
Vitamins and Minerals	
Vitamin A	172.5 IU
Vitamin B_6	0.2 mg
Vitamin B_{12}	4.6 µg
Calcium	153 mg
Iron	9.2 mg
Magnesium	51 mg
Phosphorus	493 mg
Zinc	2.9 mg
Copper	0.8 mg
Manganese	0.2 mg
Selenium	76.2 µg
Retinol	172.5 µg
Riboflavin	1.5 mg
Niacin	1.9 mg
Folate	20.4 µg
Amino Acids	
Tryptophan	0.3 g
Threonine	1.2 g
Isoleucine	1.2 g
Leucine	1.9 g
Lysine	2.1 g
Methionine	0.6 g
Cystine	0.4 g
Phenylalanine	1 g
Tyrosine	0.9 g
Valine	1.2 g
Arginine	2 g
Histidine	0.5 g
Alanine	1.7 g
Aspartic acid	2.7 g
Glutamic acid	3.8 g
Glycine	1.7 g
Proline	1.1 g
Serine	1.2 g

Source: Data from SELFNutrition Data (2014).

OCTOPUS

Enteroctopus dofleini (Wülker, 1910)

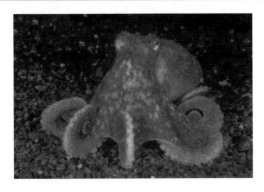

Order: Octopoda

Family: Octopodidae

Common name: North Pacific giant octopus

Distribution: Coastal North Pacific

Habitat: Much shallower or much deeper waters

Description: This species grows to a maximum weight of 272 kg with a 9 m arm span. Mantle is spherical in shape. This species can change the color of its skin, giving it the ability to blend into the environment. It is also able to change its skin texture, providing even better camouflage. Dermal muscles of the skin can create a heavily textured look through papillation or can make skin appear smooth.

Octopus ocellatus Gray, 1849

Common name: Ocellated octopus

Distribution: Pacific Ocean, including New Zealand

Habitat: Off the coasts

Description: This is a small species that grows to a maximum length of only 20 cm. Arms of this species are about two or three times its mantle length with deep lateral webs and very shallow dorsal webs. Female octopus carrying eggs in its stomach is said to be good for eating.

Nutritional Facts (Octopus) (for a Serving Size of 85 g)

Nutrients	Amount
Total fat	1.8 g
Saturated fat	0.4 g
Monounsaturated fat	0.3 g
Polyunsaturated fat	0.4 g
Cholesterol	81.6 mg
Sodium	391 mg
Potassium	535.5 mg
Carbohydrates	3.7 g
Protein	25.3 g
Vitamins and Minerals	
Vitamin A	76.5 IU
Vitamin B_6	0.6 mg
Vitamin B_{12}	30.6 µg
Vitamin C	6.8 mg
Vitamin E	1 mg
Vitamin K	0.1 µg
Calcium	90.1 mg
Iron	8.1 mg
Magnesium	51 mg
Phosphorus	237.2 mg
Zinc	2.9 mg
Copper	0.6 mg

Continued

Nutrients	Amount
Selenium	76.2 µg
Retinol	76.5 µg
Riboflavin	0.1 mg
Niacin	3.2 mg
Folate	20.4 µg
Choline	68.9 mg
Amino Acids	
Tryptophan	0.3 g
Threonine	1.1 g
Isoleucine	1.1 g
Leucine	1.8 g
Lysine	1.9 g

Methionine	0.6 g
Cystine	0.3 g
Phenylalanine	0.9 g
Tyrosine	0.8 g
Valine	1.1 g
Arginine	1.8 g
Histidine	0.5 g
Alanine	1.5 g
Aspartic acid	2.4 g
Glutamic acid	3.4 g
Glycine	1.6 g
Proline	1 g
Serine	1.1 g

Source: Data from SELFNutrition Data (2014).

7
ECHINODERMS

STARFISH

Asterias amurensis Lütken, 1871

Phylum: Echinodermata

Class: Asteroidea

Order: Decapoda

Family: Nephropidae

Common name: Sea star

Distribution: Native to the coasts of northern China, Korea, Russia, and Japan; introduced to the oceanic areas of Tasmania, southern Australia, Alaska, the Aleutian Islands, parts of Europe, and Maine

Habitat: Estuarine and marine habitats

Description: This species can grow up to 50 cm in diameter. It is yellow with red and purple pigmentation on its five arms, and a small central disk. Its distinctive characteristic is its upturned tips. The undersides are completely yellow, and the arms are unevenly covered with small, jagged-edged spines. These spines line the groove in which the tube feet lie, and they join up at the mouth in a fanlike shape.

Nutritional Facts

Protein	15.46% (weight, fresh)
Fatty Acids	
Total lipid	10.18% (wet weight basis)
Saturated fatty acids	23.1% (of total fatty acids)
Monounsaturated fatty acids	43%
Polyunsaturated fatty acids	36.5%

Source: Data from Azad Shah et al. (2013).

SEA URCHINS

Tripneustes gratilla (Linnaeus, 1758)

Phylum: Echinodermata

Class: Echinoidea

Order: Temnopleuroida

Family: Toxopneustidae

Common name: Collector urchin

Distribution: Indo-Pacific, Hawaii, and the Red Sea; from Mozambique to the Red Sea; west coast of Australia

Habitat: Open sea bottoms with some cover; depth range of 2 to 30 m

Description: Collector urchins are dark in color, usually bluish-purple with white spines. The pedicles are also white, with a dark or black base. They have orange-tipped spines. The spines of some specimens are wholly orange, whereas those of others are only orange-tipped or completely white. They reach 10 to 15 cm in size.

Nutritional Facts

Proximate Composition (of Gonad) (%)

Moisture	Ash	Protein	Lipid	Carbohydrate	Energy (Kcal/100 g)
80.7	2.8	11.7	2.8	2.0	100.8

Free Amino Acids and Dipeptides (of Gonad) (mg/100 g)

Taurine	20.6
Aspartic acid	3.1
Threonine	3.6
Serine	7.3
Glutamic acid	121.5
Sarcosine	0.1
Glycine	215.6
Alanine	120.2
Citrulline	1.1
Histidine	2.4
Valine	19.5
Cystine	0.4
Methionine	62.1
Isoleucine	2.0
Leucine	2.4
Tyrosine	6.0

Phenylalanine	3.1
Proline	48.4
Tryptophan	1.8
Arginine	41.8
Ornithine	0.8
Lysine	20.9
Cystathionine	0.5
Anserine	0.5
Carnosine	31.5
Urea	2.0
Phosphoserine	7.8
Aminoadipic acid	0.1
Aminobutyric acid	1.5
Ammonia	2.1
B-alanine	1.0
Aminoisobutyric	0.4
3-Methylhistidine	1.8

Source: Data from Chen et al. (2013).

Strongylocentrotus droebachiensis (O.F. Müller, 1776)

Order: Echinoida

Family: Strongylocentrotidae

Common name: Green sea urchin

Distribution: Northern waters all around the world, including both the Pacific and Atlantic Oceans

Habitat: Rocky substratum in the intertidal zone and up to depths of 1150 m

Description: This species is in the shape of a slightly flattened globe (dorsoventrally). The oral side rests against the substratum, and the aboral side (the side with the anus) is in the opposite direction. It has pentameric symmetry, which is visible in the five paired rows of podia (tube feet) that run from the anus to the mouth. The average adult size is around 50 mm, but it has been recorded at a diameter of 87 mm.

Nutritional Facts

Proximate Composition and SFA (Gonads and Coelomic Fluid) (%)

Moisture-Saturated Fatty Acids (SFAs)

Gonads	74.7	14.0
Coelomic fluid	96.5	16.0

Source: Data from Pathirana et al. (2002).

Strongylocentrotus franciscanus (A. Agassiz, 1863)

Order: Echinoida

Family: Strongylocentrotidae

Common name: Red sea urchin

Distribution: Gulf of Alaska to Isla Cedros, Baja California; northern Japan

Habitat: Rocky reefs, especially around kelp; low intertidal to 125 m; mostly subtidal

Description: Body is 12 cm across. Spines are up to 7 cm long. Color of spines may be red, brick red, pink, purple, or even maroon. The oldest specimens have been reported to possess a test diameter of 19 cm. Tube feet are dark, often wine red. Usually, this species eats red or brown algae, periwinkles, and occasionally barnacles or mussels.

130

Nutritional Facts (per 100 g)

Calories	172 Kcal
Protein	13.27
Polyunsaturated fatty acids	1.75
Omega-3 fatty acids	1.07

Eicosapentaenoic acid (EPA)	0.79
Docosahexaenoic acid (DHA)	0.04
Zinc	17.00 ppm

Source: Data from Pacific Urchin Harvesters Association (2013).

SEA CUCUMBERS

Acaudina molpadioides (Semper, 1867)

Phylum: Echinodermata

Class: Holothuroidea

Order: Molpadida

Family: Caudinidae

Common name: Sweet sea potato

Distribution: Andaman and Nicobar Islands, East China Sea, Eastern India, Eastern Philippines, Gulf of Tonkin, Halmahera, Northern Bay of Bengal, South India, Sri Lanka, Southern China, Western India, and Yellow Sea

Habitat: Reef, estuarine, and muddy areas; depths of about 20–50 m

Description: This species measures about 15–20 cm in length. Its body is consistently flesh colored. The body is stout and spindle shaped, tapering at the posterior end. An additional feature is the presence of 15 tentacles on its body. The body wall is smooth and opaque. Tube feet are absent.

Thelenota ananas Jaeger, 1833

Order: Aspidochirotida

Family: Stichopodidae

Common name: Prickly redfish

Distribution: Throughout the Indo-Pacific, excluding Hawaii

Habitat: Slopes and passes within reef zones; outer reef flats; depths of 35 m

Description: It is a gray, orange, and/or red sea cucumber, often with a purple cast. It is a very large species (750 mm), square in cross-section with prominent "cockscomb" papillae over its upper surface. It has a thick but pliable body and a smooth tegument. Spicules are delicate, dichotomously branched rods without lateral spines.

Nutritional Facts

Proximate Composition

Item	Protein (%)	Fat (%)	Moisture (%)	Carbohydrate (%)	Ash (%)
Fresh body wall of AM	11.52	0.03	87.83	0.38	0.99
Dried AM	68.53	0.55	8.25	—	7.56
Fresh body wall of TA	16.64	0.27	76.97	2.47	1.60

AM: *Acaudina molpadioides*; TA: *Thelenota ananas*.
Source: Data from Bordbar et al. (2011).

Apostichopus japonicus (Selenka, 1867)

Order: Aspidochirotida

Family: Stichopodidae

Common name: Japanese spiky sea cucumber

Distribution: Northwest Pacific, including Japan, China, Korean Peninsula, and far eastern Russia

Habitat: Shallow coastal bottom communities from the intertidal zone to depths of 40 m

Description: It has a cylindrical leathery body with blunt, thorny protuberances. At the anterior or front end, there is a mouth surrounded by a ring of short feeding tentacles, and at the posterior end is the anus. There are three different color morphs: red, green, and black.

Nutritional Facts

Proximate Composition

Moisture (%)	Ash (%)	Protein (%)	Lipid (%)
80.26	2.57	1.13	0.14 (min.)
91.49	6.85	3.99	2.12 (max.)

Source: Data from Lee (2012).

Holothuria (Metriatyla) scabra Jaeger, 1833

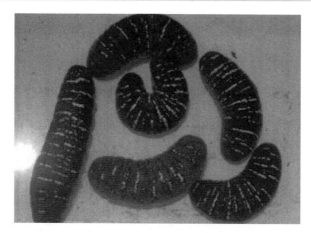

Order: Aspidochirotida

Family: Holothuriidae

Common name: Sandfish

Distribution: Indo-Pacific region

Habitat: Reef flats, slopes, and shallow seagrass beds

Description: It is grayish-black on the upper side with dark-colored wrinkles but paler on the underside. It grows up to 4 cm long and is broader than high. It has a pliable skin and is covered by calcareous spicules in the form of tablets and buttons.

Holothuria (Microthele) nobilis (Selenka, 1867)

Order: Aspidochirotida

Family: Holothuriidae

Common name: Black teatfish

Distribution: Africa and Indian Ocean region

Habitat: Coral reef habitat, reef flats, and outer slopes

133

Description: It is a black sea cucumber, often with a fine sand covering. It is a large, stout species, firm and rigid, with prominent lateral papillae and anal papillae (usually five). It has translucent cuvierian tubules. Spicules are stout tablets and fenestrated ellipsoids.

Holothuria (Thymiosycia) impatiens (Forskål, 1775)

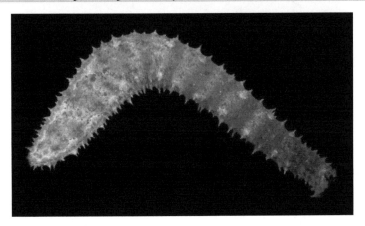

Order: Aspidochirotida

Family: Holothuriidae

Common name: Sea cucumber

Distribution: Belize, Caribbean Sea, Colombia, Cuba, Djibouti, Federal Republic of Somalia, Gulf of Mexico, Indian Ocean, Jamaica, Kenya, Madagascar, Mediterranean Sea, Mozambique, Panama, Red Sea, Republic of Mauritius, Seychelles Islands, South Africa, Tanzania, and Venezuela

Habitat: Near-shore habitats; on reefs, below rocks on the reef flat

Description: This species has a variegated pink and brown coloring. It is a small to large (150–400 mm), cylindrical species with a tough tegument, prominent papillae, and a firm but pliable body. Cuvierian tubules may sometimes be ejected. Spicules are well-developed square tablets and smooth buttons.

Holothuria pardalis Selenka, 1867 (= *Holothuria (Lessonothuria) insignis*)

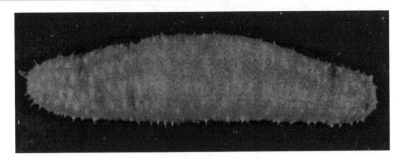

Order: Aspidochirotida

Family: Holothuriidae

Common name: Leopard sea cucumber

Distribution: From Madagascar and East Africa to the Red Sea, from Australia to China, and east across the Pacific to Hawaii, Cocos Island, Costa Rica, the Galapagos Islands, Ecuador, and Colombia

Habitat: Shallow waters (between 0 and 10 m) over blocks of coral or buried in the coral rubble

Description: It is a mottled, light brown–gray and white sea cucumber, sometimes with dark spots along its upper surface. It is cylindrical and tapering at its ends. It is a small species (<100 mm), with a smooth tegument and a thin, pliable body. Spicules are clumsy tablets with a low to moderate spire and a spiny disc, and buttons are usually irregular.

Holothuria (Lessonothuria) multipilula Liao, 1975

Order: Aspidochirotida

Family: Holothuriidae

Common name: Holothuria

Distribution: China

Habitat: Shallow waters (between 0 and 10 m)

Description: This species has a globoid body with meriodional symmetry (i.e., no arms). Brachioles or appendages or pinnate structure is absent. Larvae have bilateral symmetry, but this is lost during metamorphosis.

Actinopyga echinites (Jaeger, 1833)

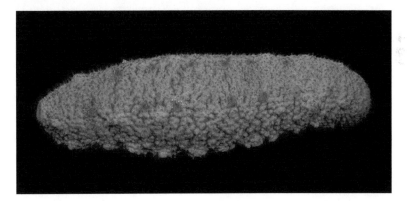

Order: Aspidochirotida

Family: Holothuriidae

Common name: Deepwater redfish

Distribution: Found throughout the Western Central Pacific, Asia, Africa, and Indian Ocean regions

Habitat: Outer reef flats in the littoral zone, and in estuaries and lagoons

Description: It is a brown sea cucumber with a rough tegument with numerous papillae (size: 300–350 mm). The upper surface is often covered with sand. The spicules include large rods.

Nutritional Facts

Amino Acids	A	B	C	D	E	F	G
Nonessential Amino Acids (mg/g)							
Aspartine	3.69	6.59	5.20	3.26	3.50	4.84	5.78
Serine	1.31	2.91	2.53	1.33	1.48	2.16	2.07
Glutamic acid	6.43	11.13	9.82	5.72	6.75	8.30	7.86
Proline	3.08	3.32	4.57	2.40	3.35	4.11	1.03
Glycine	8.09	17.08	10.02	4.50	7.32	8.43	10.03
Alanine	4.10	8.41	5.54	2.69	4.10	4.80	5.02
Cysteine	0.46	—	0.49	0.52	—	1.17	—
Tyrosine	0.99	1.65	1.55	1.12	1.06	1.70	1.41
Phenylalanine	1.15	1.45	1.78	1.40	1.12	1.99	1.67
Subtotal	29.30	52.54	41.50	22.94	28.68	37.50	34.87
Essential Amino Acids (mg/g)							
Lysine	0.64	1.02	1.59	0.38	1.09	1.45	0.92
Histidine	0.17	0.37	0.45	2.82	0.24	0.42	0.40
Arginine	3.40	6.60	4.95	1.63	3.45	4.23	4.46
Valine	1.59	2.64	2.23	1.09	1.78	2.50	2.43
Methionine	0.89	1.03	1.40	1.21	0.90	1.49	0.86
Isoleucine	0.76	1.39	1.45	1.93	0.98	1.51	1.64
Leucine	1.49	2.64	2.64	1.74	1.75	2.63	2.59
Threonine	1.68	3.44	2.68	—	1.89	2.48	2.58
Subtotal	10.62	19.13	17.39	10.80	12.08	16.75	15.88
Total	39.92	71.67	58.89	33.74	40.76	54.25	50.75

A: *H. (Metriatyla) scabra*; B: *H. (Microthele) nobilis*; C: *H. (Thymiosycia) impatiens*; D: *H. (Lessonothuria) insignis*; E: *H. (Lessonothuria) multipilula*; F: *Actinopyga echinites*; G: *Thelenota ananas*.

Minerals (mg/100g)

Element	A	B	C	D	E	F	G
Ba	2.0	6.4	1.7	1	2.9	3.6	2
Co	0.4	—	0.2	0.4	1.7	0.7	0.4
Cr	10.1	12.9	15.3	9.3	4.4	11.5	10.1
Cu	6.1	1.3	5.9	2	2.5	1.8	6.1
Li	2.0	1.2	1.7	1	1.9	1.4	2
Mn	19.1	2.6	4.1	5.8	11.6	36.1	4.1
Ni	2.9	2.5	5	2.3	1.9	2.1	5
Si	110	12.9	170	11.5	77.6	65	46.6
Sr	616	181	119	57.8	874	64	162
V	—	0.51	0.51	0.34	0.97	0.72	1.01
Zn	28.6	111	40.9	10.4	9.71	26	70.9

A: *H. (Metriatyla) scabra*; B: *H. (Microthele) nobilis*; C: *H. (Thymiosycia) impatiens*; D: *H. (Lessonothuria) insignis*; E: *H. (Lessonothuria) multipilula*; F: *Actinopyga echinites*; G: *Thelenota ananas*.

8
PROCHORDATE

Halocynthia roretz (VonDrasche, 1884)

Phylum: Chordata

Class: Ascidiacea

Order: Pleurogona

Family: Pyuridae

Common name: Sea pineapple

Distribution: Ubiquitous throughout the world

Habitat: Subtidal rocky bottom

Description: The species of sea squirts are somewhat barrel-shaped animals and the entire body is invested with a thick covering, the tunic. An individual has two openings: an incurrent branchial (oral) siphon and an outcurrent atrial siphon. Ascidians live by filtering tiny plankton (diatoms, protozoans, copepods, and larvae of various invertebrates) and nutrient materials from seawater. The animals are hermaphrodite.

Nutritional Facts

Proximate Composition (g per 46 g)

Energy	Protein	Fat	Carbohydrate
14*	2.3	0.37	0.37

* Calories.

Minerals (mg per 46 g)

Na	K	Ca	Mg	P	Fe	Zn	Cu
598	262	14.7	18.9	25.3	2.6	2.4	0.1

Vitamins (mg per 46 g)

B$_2$	Niacin	B$_6$	B$_{12}$*	Folate*	Pantothenic Acid	C
0.06	0.23	0.01	1.75	14.7	0.15	1.38

* Micrograms.

Herdmania pallida (Heller, 1878)

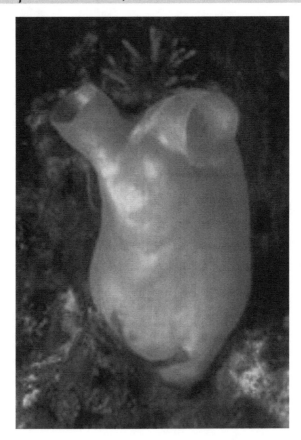

Phylum: Chordata

Class: Ascidiacea

Order: Pleurogona

Family: Pyuridae

Common name: NA

Distribution: Djibouti, Gulf of Mexico

Habitat: Subtidal rocky bottom

Description: It is potato-like in shape and pink in color. On the free side, the body shows two projections, the branchial and atrial siphons. The branchial siphon is short. The branchial siphon shows a branchial aperture or mouth. The atrial siphon is longer. It bears the atrial aperture. Both openings are bounded by four lips. The body of this animal is covered and protected by test, a thick, leathery covering of the body that is secreted by the epidermis of the body wall.

Nutritional Facts

Proximate Composition (mg/100 mg Dry Weight)

Protein	Lipid	Carbohydrate	Amino Acids
45	18	4.8	1.201

Source: Data from Tamilselvi et al. (2010).

9
FISH

TELEOSTS

Chirocentrus dorab (Forsskål, 1775)

Phylum: Chordata

Class: Actinopterygii

Order: Clupeiformes

Family: Chirocentridae

Common name: Wolf herring

Distribution: Indo-Pacific

Habitat: Marine; brackish; reef-associated; amphidromous

Description: This species has a maximum length of 1 m. Scales are numerous, small, and usually lost. Body is silvery; back is bright blue (fading to gray); and flanks are bright silver. The slightly shorter pectoral fin and the black marking of the upper part of the dorsal fin are the chief identifying features of this species. Fins are spineless.

Nutritional Facts

Proximate Composition (per 100 g of Edible Portion)

Energy (Kcal)	Moisture (%)	Protein (g)	Fat (g)	Carbohydrate (g)	Ash (g)
101	75.9	21.8	1.5	0	1.4

Minerals (mg)

Ca	P	Fe	Na	K	Mg
23	246	0.5	6.7	276	1314

Source: Data from Siong et al. (1987).

Co	Cu	Mn	Zn
0.40	98.85	59.51	340.78

Source: Data from Nurnadia et al. (2013).

Vitamins (mg)

Thiamine	Riboflavin	Niacin	Ascorbic Acid
0.02	0.05	3.4	0.9

Source: Data from Siong et al. (1987).

Dussumieria acuta Valenciennes, 1847

Order: Clupeiformes

Family: Dussumieriidae

Common name: Rainbow sardine

Distribution: Indo-Pacific

Habitat: Marine; freshwater; brackish; pelagic-neritic; depth range 10–20 m

Description: This species has a maximum length of 20 cm. Color of the body is iridescent blue with a shiny gold or brass line below. Hind margin of the tail is broadly dark. W-shaped pelvic scute is present. Isthmus is tapering evenly forward, and anal fin rays are many.

Ilisha melanostoma (Schneider, 1801)

Order: Clupeiformes

Family: Dussumieriidae

Common name: Indian ilisha

Distribution: Indian Ocean

Habitat: Marine; pelagic; coastal, but entering estuaries

Description: It grows to a maximum size of 17 cm (standard length). Body is moderately deep. Belly has a total of 25 to 30 scutes. Eye is large, and lower jaw is projecting. Dorsal fin originates a little before the midpoint of the body. Anal fin has 35 to 48 finrays. Swimbladder is with two tubes passing back in the muscles on either side of the haemal spines.

Pellona ditchela Valenciennes, 1847

Order: Clupeiformes

Family: Pristigasteridae

Common name: Indian pellona

Distribution: Indo-West Pacific

Habitat: Marine; freshwater; brackish; pelagic-neritic; anadromous

Description: It grows to a maximum length of 16 cm. Belly has a total 26 to 28 scutes and is strongly keeled. Eye is large, and lower jaw is projecting. Upper jaw has a toothed hypo-maxillary bone between hind tip of pre-maxilla and lower bulge of maxilla blade. Dorsal fin originates near midpoint of the body.

Sardinella longiceps Valenciennes, 1847

Order: Clupeiformes

Family: Clupeidae

Common name: Indian oil sardine

Distribution: Indian Ocean

Habitat: Pelagic-neritic; oceanodromous; depth range 20–200 m

Description: This species has a maximum length of 23 cm and weight of 200 g. Body is subcylindrical. A faint golden spot is seen behind gill opening which is followed by a faint golden midlateral line. A distinct black spot is present at the hind border of gill cover. This species is largely distinguished by its longer head and more lower gill rakers. There is no prominent keel.

Sardinella fimbriata (Valenciennes, 1847)

Order: Clupeiformes

Family: Clupeidae

Common name: Fringe scale sardinella

Distribution: Indo-West Pacific

Habitat: Marine; brackish; pelagic-neritic; depth range 0–50 m

Description: This species has a maximum length of 13 cm. Body is somewhat compressed but variable. Total number of scutes varies from 29 to 33. Hind part of scales has a few perforations, and they are somewhat produced posteriorly. A dark spot is seen at the dorsal fin origin.

Nutritional Facts

Proximate Composition

Species	Moisture (%)	Protein (%)	Fat (%)	Ash (%)	Carbohydrate (%)	Energy Value (Kcal/100 g)
Dussumieria acuta	74.07	20.23	6.83	1.22	0.083	150.72
Ilisha melanostoma	75.14	15.33	4.06	1.16	0.158	98.49
Pellona ditchela	78.03	15.38	0.37	1.21	0.177	65.56
Sardinella longiceps	73.79	15.94	6.00	1.23	0.358	116.95
Sardinella fimbriata	76.49	18.16	3.20	1.37	0.100	101.82

Minerals (mg/100 g)

	Ca	Fe	P
Dussumieria acuta	879.97	14.18	124.35
Ilisha melanostoma	846.20	2.06	413.91
Pellona ditchela	890.38	2.96	226.19
Sardinella longiceps	1026.85	4.31	352.6
Sardinella fimbriata	164.29	2.25	346.83

Source: Data from Palanikumar et al. (2014).

Minerals (mg/100 g)

Na	K	Ca	Mg	Co	Cu	Mn	Zn
0.54	51.35	23.67	387.71	23.34	11.22	78.38	669.66

Source: Data from Nurnadia et al. (2013).

Ilisha elongata (Anonymous [Bennett], 1830)

Order: Clupeiformes

Family: Pristigasteridae

Common name: Slender shad

Distribution: Indo-Pacific

Habitat: Marine; brackish; pelagic-neritic

Description: This species has a maximum length of 60 cm and weight of 140 g. Body is slender, and belly has a total of 34 to 42 scutes. Eye is large, and lower jaw is projecting. Dorsal fin originates at about midpoint of the body or a little behind, and anal fin originates a little behind the dorsal fin base.

Swim bladder has a long tube passing back down the right side of the body above the anal fin base.

Nutritional Facts

Proximate Composition (per 100 g of Edible Portion)

Energy (Kcal)	Moisture (%)	Protein (g)	Fat (g)	Carbohydrate (g)	Ash (g)
85	78.7	18.9	1.0	0	1.4

Minerals (mg)

Ca	P	Fe	Na	K
56	195	0.4	183	314

Vitamins (mg)

Thiamine	Riboflavin	Niacin	Ascorbic Acid
0	0.05	2.5	0

Source: Data from Siong et al. (1987).

Tenualosa macrura (Bleeker, 1852) (= *Hilsa (Clupea) macrura*)

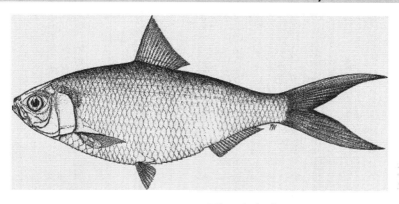

Order: Clupeiformes

Family: Clupeidae

Common name: Longtail shad

Distribution: Western Central Pacific

Habitat: Marine; freshwater; brackish; pelagic-neritic; anadromous

Description: This species has a maximum length of 52 cm. Body is moderately deep, and the belly has 30 scutes. A median notch is seen on the upper jaw. Gill rakers are fine but not numerous. Caudal fin is long, and the lobes are long and pointed.

Minerals (mg)

Ca	P	Fe	Na	K	Mg
35	198	1.1	67	227	925

Source: Data from Siong et al. (1987).

Co	Cu	Mn	Zn
0.87	40.72	53.81	413.57

Source: Data from Nurnadia et al. (2013).

Vitamins (mg)

Thiamine	Riboflavin	Niacin	Ascorbic Acid
0	0.17	2.5	0

Source: Data from Siong et al. (1987).

Nutritional Facts

Proximate Composition (per 100 g of Edible Portion)

Energy (Kcal)	Moisture (%)	Protein (g)	Fat (g)	Carbohydrate (g)	Ash (g)
281	59.4	18.4	2.3	0	1.1

Dussumieria hasselti Bleeker, 1849

Order: Clupeiformes

Family: Clupeidae

Common name: Round herring

Distribution: Restricted to the tropical and subtropical portions of the Indian Ocean and the South China Sea

Habitat: Pelagic—found near shores

Description: It grows to 8–18 cm. Color of the body is silvery on the sides, darkening to bluish gray on the back and lightening to almost white on the underside. It has an elongated body, with a round belly and a pointed nose. The single dorsal (back) fin is slightly behind midpoint. The tail fin is heavily forked. The scales are very delicate and are easily detached. Like other sardines, it has no lateral line and no scales on the head.

Nutritional Facts

Proximate Composition (per 100 g of Edible Portion)

Energy (Kcal)	Moisture (%)	Protein (g)	Fat (g)	Carbohydrate (g)	Ash (g)
118	73.7	21.5	3.5	0	1.4

Minerals (mg)

Ca	P	Fe	Na	K
0	274	1.3	97	318

Source: Data from Siong et al. (1987).

Vitamins (mg)

Thiamine	Riboflavin	Niacin	Ascorbic Acid
0.13	0.29	4.2	0

Source: Data from Siong et al. (1987).

Clupea harengus Linnaeus, 1758

Order: Clupeiformes

Family: Clupeidae

Common name: Atlantic herring

Distribution: North Atlantic

Habitat: Marine; brackish; benthope-lagic; oceanodromous; depth range 0–364 m

Description: This species has a maximum length of 45 cm and weight of 1.1 kg. It is a slender fish with a round belly. Scutes are without a prominent keel. About 15 postpelvic scutes are present. Body is blue to greenish-blue dorsally and silvery ventrally. Distinctive dark spots are absent on the body or fins.

Nutritional Facts (per 100 g)

Water	72.050 g
Energy	158.000 Kcal
Energy	661.000 kJ
Protein	17.960 g
Total lipid (fat)	9.040 g
Ash	1.460 g
Carbohydrate by difference	—
Dietary fiber	—

Sugars, total	—
Calcium, Ca	57.000 mg
Iron, Fe	1.100 mg
Magnesium, Mg	32.000 mg
Phosphorus, P	236.000 mg
Potassium, K	327.000 mg
Sodium, Na	90.000 mg
Zinc, Zn	0.990 mg
Copper, Cu	0.092 mg
Manganese, Mn	0.035 mg
Selenium, Se	36.500 µg
Vitamin C, total ascorbic acid	0.700 mg
Thiamin	0.092 mg
Riboflavin	0.233 mg
Niacin	3.217 mg
Pantothenic acid	0.645 mg
Vitamin B-6	0.302 mg
Folate, total	10.000 µg
Folic acid	—

Source: Data from *Titi Tudorancea Bulletin* (2014).

Ethmalosa fimbriata (Bowdich, 1825)

Order: Clupeiformes

Family: Clupeidae

Common name: Bonga shad

Distribution: Eastern Central Atlantic

Habitat: Marine; freshwater; brackish; pelagic-neritic; catadromous

Description: This species has a maximum length of 45 cm. Upper jaw has distinct median notch, into which tip of lower jaw fits. Lower gill rakers are long, fine, and numerous, about three times as long as gill filaments. Upper gill rakers are bent sharply upward and V-shaped. Caudal fin is deep chrome, and tips are long and pointed. Dorsal fin tip is black. Golden tints are present on the body.

Nutritional Facts

Proximate Composition

Energy (kJ/kg)	Moisture (%)	Protein (g)	Fat (g)	Carbohydrate (g)	Ash (g)
66.50	18.50	1.90	8.70	9.63	449.55

Minerals (mg/100 g)

Na	K	Ca	Mg	Fe	Zn	Cu	P	Mn	Pb
307.8	11.4	92.8	7.3	114.5	1.2	3.4	6.9	33.5	4.2

Source: Data from Udo and Arazu (2012).

Fatty Acids

Fat (g/100 g)	3.06
Cholesterol (mg/100 g)	41.8

Saturated fatty acid	9.9 (%)
Monounsaturated	9.1 (%)
Polyunsaturated (ω6)	11.0 (%)
(ω3)	30.9 (%)
Other polyunsaturated	35.0 (%)
Total fatty acids	76.9 (%)

Source: Data from Osman et al. (2001).

Anadontosoma chacunda (Hamilton, 1822)

Order: Clupeiformes

Family: Clupeidae

Common name: Gizzard shad

Distribution: Indo-West Pacific

Habitat: Marine; freshwater; brackish; pelagic-neritic; anadromous; depth range 0–50 m

Description: This species grows to a maximum length of 22 cm. Body depth increases with size of fish. Longest gill rakers on the lower part of the arch are less than corresponding gill filaments. Hind edges of scales are toothed. A median series of pre-dorsal scales are present. A large black spot is seen behind the gill opening.

Nutritional Facts

Proximate Composition (per 100 g of Edible Portion)

Energy (Kcal)	Moisture (%)	Protein (g)	Fat (g)	Carbohydrate (g)	Ash (g)
132	74.1	19.7	5.9	0	1.7

Minerals (mg)

Ca	P	Fe	Na	K
105	260	0.8	124	358

Vitamins (mg)

Thiamine	Riboflavin	Niacin	Ascorbic Acid
0	0.26	2.1	0

Source: Data from Siong et al. (1987).

Stolephorus commersonii **Lacepède, 1803**

Order: Clupeiformes

Family: Engraulidae

Common name: Anchovy

Distribution: Indo-West Pacific

Habitat: Marine; brackish; pelagic-neritic; anadromous; depth range 0–50 m

Description: This species grows to a maximum length of 10 cm. Belly is slightly rounded with 0–5 small, needle-like pre-pelvic scutes. Small teeth are seen on hyoid bones. Isthmus muscle tapers evenly forward. Body color is light transparent, fleshy brown with a pair of dark patches behind the occiput, followed by a pair of lines to the dorsal fin origin. It bears a silver stripe on the flanks.

Nutritional Facts

Proximate Composition (per 100 g of Edible Portion)

Energy (Kcal)	Moisture (%)	Protein (g)	Fat (g)	Carbohydrate (g)	Ash (g)
74	80.6	17.7	0.3	0	1.7

Minerals (mg)

Ca	P	Fe	Na	K
134	162	0.6	573	130

Vitamins (mg)

Thiamine	Riboflavin	Niacin	Ascorbic Acid
0.02	0.09	1.2	0.8

Source: Data from Siong et al. (1987).

Fatty Acids

EPA (µg/g)	DHA (µg/g)
45.1	117.2

EPA: eicosapentaenoic acid; DHA: docosahexaenoic acid.

Source: Data from Wan Rosli et al. (2012).

Cottoperca gobio **(Gunther, 1861)**

Order: Perciformes

Family: Bovichtidae

Common name: Channel Bull Blenny

Distribution: Southeast Pacific and Southwest Atlantic

Habitat: Demersal

Description: This species has a maximum length of 80.0 cm with dorsal soft rays and anal soft rays numbering 23 and 21, respectively

Nutritional Facts

Proximate Composition

Moisture (%)	Ash (%)	Lipid (%)	Protein (%)	Energetic Value (kJ/g Wet Mass)
74.78	3.36	2.49	13.93	4.28

Values based on percentage of wet mass.
Source: Data from Eder and Lewis (2005).

Brama brama (Bonnaterre, 1788)

Order: Perciformes

Family: Bramidae

Common name: Atlantic pomfret, Ray's bream

Distribution: Atlantic, Indian, and South Pacific Oceans; Western Atlantic and Eastern Atlantic Oceans

Habitat: Bathypelagic; oceanodromous; depth range 0–1000 m

Description: This species grows to a maximum length and size of 1 m and 6 kg. It possesses a compressed, deep body with a steeply curved head profile. Dorsal and anal fins are scaled and have rigid fin rays.

Nutritional Facts

Proximate Composition

Moisture (%)	Ash (%)	Lipid (%)	Protein (%)	Energetic Value (kJ/g Wet Mass)
67.66	2.09	10.85	18.38	8.65

Values based on percentage of wet mass.
Source: Data from Eder and Lewis (2005).

150

Caesio erythrogaster (Bloch, 1791)

Order: Perciformes

Family: Caesionidae

Common name: Redbelly yellowtail fusilier

Distribution: Indo-West Pacific

Habitat: Reef associated; nonmigratory; depth range 1–60 m.

Description: This species has a maximum length of 60 cm with a deep body. Dorsal and anal fins are scaled. Posterior end of the maxilla is blunt. Coloration of the upper body is grayish blue; lower sides and belly are white or pinkish. Pectoral, pelvic, and anal fins are white to pink. Large yellow tail is present. Dorsal fin is yellow posteriorly and grayish blue anteriorly.

Nutritional Facts

Proximate Composition (per 100 g of Edible Portion)

Energy (Kcal)	Moisture (%)	Protein (g)	Fat (g)	Carbohydrate (g)	Ash (g)
92	77.4	21.5	0.8	0	1.4

Minerals (mg)

Ca	P	Fe	Na	K
82	235	0.4	47	487

Vitamins (mg)

Thiamine	Riboflavin	Niacin	Ascorbic Acid
0.05	0.16	2.8	0

Source: Data from Siong et al. (1987).

Trachinotus blochii (Lacepède, 1801)

Order: Perciformes

Family: Carangidae

Common name: Snubnose pomp

Distribution: Indo-Pacific

Habitat: Brackish; reef-associated

Description: This species has a maximum length and weight of 110 cm and 3 kg. Dorsal soft rays and anal soft rays number 20 and 18, respectively. Juveniles are in small schools, while adults are usually solitary.

Nutritional Facts

Proximate Composition (per 100 g of Edible Portion)

Energy (Kcal)	Moisture (%)	Protein (g)	Fat (g)	Carbohydrate (g)	Ash (g)
95	76.7	21.7	0.9	0	1.3

Minerals (mg)

Ca	P	Fe	Na	K
41	234	0.7	85	455

Vitamins (mg)

Thiamine	Riboflavin	Niacin	Ascorbic Acid
0.06	0.16	3.3	0

Source: Data from Siong et al. (1987).

Parona signata (Jenyns, 1841)

Order: Perciformes

Family: Carangidae

Common name: Parona leatherjacket

Distribution: Southwest Atlantic

Habitat: Demersal

Description: This species has a maximum length of 60 cm. The first dorsal fin is composed of several short spines. Two detached anal spines are anterior to the soft anal fin. The lateral line is curved and undulated, with more or less 10 branches anteriorly and fairly straight posteriorly. A black patch is seen on the opercle, which is a conspicuous character of this species.

Nutritional Facts

Proximate Composition

Moisture (%)	Ash (%)	Lipid (%)	Protein (%)	Energetic Value (kJ/g Wet Mass)
62.43	2.49	16.24	16.23	10.27

Values based on percentage of wet mass.
Source: Data from Eder and Lewis (2005).

Seriola lalandi Valenciennes, 1833

Order: Perciformes

Family: Carangidae

Common name: Yellowtail amberjack, great amberjack

Distribution: Circumglobal in sub-tropical waters: Indo-Pacific, Eastern Pacific, and Eastern Atlantic

Habitat: Brackish; benthopelagic; depth range 3–825 m

Description: This species has a maximum length of 250 cm and weight of 96 kg. This is the only jack without scutella on the caudal peduncle. Body is dark blue dorsally and almost white ventrally; with a well-defined line of demarcation between the two colors.

Nutritional Facts

Proximate Composition (Dry Weight Basis)

Protein (g)	Carbohydrate (g)	Lipid (g)	Ash (g)
12.5	4.2	3.1	1.1

Vitamins (mg/100 g)

B_1	B_2	B_3	B_5	B_6	FA	A	K	D_3
0.56	0.67	0.49	0.71	0.41	0.81	0.19	0.65	0.37

FA: folic acid.

Source: Data from Dhaneesh et al. (2012).

Parastromateus niger (Bloch, 1795)

Order: Perciformes

Family: Carangidae

Common name: Black pomfret

Distribution: Indo-West Pacific

Habitat: Marine; brackish; reef-associated; amphidromous; depth range 15–105

Description: This species grows to a maximum length of 75 cm and is a deep-bodied and strongly compressed fish. Lateral line ends in weakly developed scutes on the caudal peduncle. Pelvic fins are absent in individuals over 9 cm. Color of the body is brown above and silvery-white below. The anterior parts of the dorsal and anal fins are bluish-gray. The other fins are yellowish.

Nutritional Facts

Proximate Composition (per 100 g of Edible Portion)

Energy (Kcal)	Moisture (%)	Protein (g)	Fat (g)	Carbohydrate (g)	Ash (g)
104	75.6	20.0	2.4	0.1	1.4

Vitamins (mg)

Thiamine	Riboflavin	Niacin	Ascorbic Acid
0.28	0.18	4.2	0

Source: Data from Siong et al. (1987).

Minerals (mg)

Ca	P	Fe	Na	K	Mg	Co	Cu	Mn	Zn
0.57	54.25	15.74	288.43	39	237	0.5	57	226	944

Source: Data from Siong et al. (1987); and Nurnadia et al. (2013).

Fatty Acids

EPA (µg/g)	DHA (µg/g)
24.5	100.9

EPA: eicosapentaenoic acid; DHA: docosahexaenoic acid.

Source: Data from Wan Rosli et al. (2012).

Fat (g/100 g)	2.79
Cholesterol (mg/100 g)	46.8
Saturated fatty acid (%)	11.4
Monounsaturated (%)	3.6
Polyunsaturated (ω6) (%)	13.5
(ω3) (%)	30.5
Other polyunsaturated (%)	30.6

Source: Data from Osman et al. (2001).

Selaroides leptolepis (Cuvier, 1833)

Order: Perciformes

Family: Carangidae

Common name: Smoked yellow stripe trevally

Distribution: Indo-West Pacific

Habitat: Brackish; reef-associated; amphidromous; depth range 1–50 m

Description: This species has a maximum length of 22 cm and weight of 600 g. They form large demersal schools over soft bottom habitats. Sometimes, they ascend into freshwater reaches.

Nutritional Facts

Proximate Composition (per 100 g of Edible Portion)

Energy (Kcal)	Moisture (%)	Protein (g)	Fat (g)	Carbohydrate (g)	Ash (g)
106	75.5	21.4	2.3	0	1.4

Vitamins (mg)

Thiamine	Riboflavin	Niacin	Ascorbic Acid
0.05	0.22	3.4	0

Source: Data from Siong et al. (1987).

Minerals (mg)

Ca	P	F	Na	K	Mg
54	260	0.9	43	391	937

Source: Data from Siong et al. (1987).

Co	Cu	Mn	Zn
0.30	141.26	18.78	449.49

Source: Data from Nurnadia et al. (2013).

Fatty Acids

EPA (µg/g)	DHA (µg/g)
36.6	95.9

EPA: eicosapentaenoic acid; DHA: docosa-hexaenoic acid.
Source: Data from Wan Rosli et al. (2012).

Fat (g/100 g)	5.77
Cholesterol (mg/100 g)	47.3
Saturated fatty acid (%)	16.1
Monounsaturated (%)	11.4
Polyunsaturated (ω6) (%)	12.0
(ω3) (%)	41.3
Other polyunsaturated (%)	32.3
Total fatty acids (%)	85.5

Source: Data from Osman et al. (2001).

Megalapsis cordyla (Linnaeus, 1758)

Order: Perciformes

Family: Carangidae

Common name: Hardtail scad

Distribution: Indo-West Pacific

Habitat: Brackish; reef associated; depth range 20–100 m

Description: This species has a maximum length of 80 cm and weight of 4 kg. Dorsal soft rays and anal soft rays number 20 and 17, respectively. Adults are primarily oceanic, with pelagic schooling species rarely seen on reefs.

Nutritional Facts

Proximate Composition (per 100 g of Edible Portion)

Energy (Kcal)	Moisture (%)	Protein (g)	Fat (g)	Carbohydrate (g)	Ash (g)
105	73.7	21.0	1.0	2.9	1.4

Vitamins (mg)

Thiamine	Riboflavin	Niacin	Ascorbic Acid
0.05	0.25	3.6	0

Source: Data from Siong et al. (1987).

Minerals (mg)

Ca	P	Fe	Na	K	Mg
64	277	2.5	72	263	875

Source: Data from Siong et al. (1987).

Co	Cu	Mn	Zn
0.93	156.56	21.24	398.60

Source: Data from Nurnadia et al. (2013).

Fatty Acids

EPA (µg/g)	DHA (µg/g)
34.6	145.9

EPA: eicosapentaenoic acid; DHA: docosahexaenoic acid.

Source: Data from Wan Rosli et al. (2012).

Fat (g/100 g)	3.08
Cholesterol (mg/100 g)	46.6
Saturated fatty acid (%)	9.5
Monounsaturated (%)	3.6
Polyunsaturated (ω6) (%)	11.7
(ω3) (%)	48.4
Other polyunsaturated (%)	27.7
Total fatty acids (%)	86.9

Source: Data from Osman et al. (2001).

Carangoides malabaricus (Bloch & Schneider, 1801)

Order: Perciformes

Family: Carangidae

Common name: Malabar trevalley

Distribution: Indo-West Pacific

Habitat: Reef-associated; amphidromous; depth range 20–140 m

Description: This species has a maximum length of 60 cm. Dorsal soft rays and anal soft rays number 23 and 19, respectively. Adults feed on crustaceans, small squids, and fishes. Juveniles are found in sandy bays.

Nutritional Facts

Proximate Composition (per 100 g of Edible Portion)

Energy (Kcal)	Moisture (%)	Protein (g)	Fat (g)	Carbohydrate (g)	Ash (g)
100	76.7	20.8	1.9	0	1.1

Minerals (mg)

Ca	P	Fe	Na	K
46	187	0.6	82	217

Vitamins (mg)

Thiamine	Riboflavin	Niacin	Ascorbic Acid
0.03	0.20	3.4	0

Source: Data from Siong et al. (1987).

Carangoides orthogrammus (Jordan & Gilbert, 1882)

Order: Perciformes

Family: Carangidae

Common name: Island trevally

Distribution: Indo-Pacific

Habitat: Reef associated; oceanodromous; depth range 3–168 m

Description: This species has a maximum length of 75 cm and weight of 7 kg. Adults are pelagic and are abundant around oceanic islands while not found in neritic areas. This species may be encountered solitary, in pairs, or in small schools. Schools frequent sandy river basins, sandy channels of lagoon, and seaward reefs.

Nutritional Facts

Proximate Composition (Dry Weight Basis)

Protein (g)	Carbohydrate (g)	Lipid (g)	Ash (g)
11.5	3.9	4.2	1.5

Vitamins (mg/100 g)

B_1	B_2	B_3	B_5	B_6	FA	A	K	D_3
0.42	0.13	0.31	0.81	0.19	0.92	0.62	0.65	0.51

FA: folic acid.

Source: Data from Dhaneesh et al. (2012).

Selar mate (G. Cuvier, 1833)

Order: Perciformes

Family: Carangidae

Common name: One-finlet scad

Distribution: Tropical and subtropical regions of the Indo-Pacific region

Habitat: Coastal, schooling in inshore waters to a depth of 80 m, often in large embayments with mangroves or over coral reefs.

Description: It has a moderately compressed, oval-shaped body. The dorsal and ventral profiles of the fish are nearly evenly convex, with the two lines intersecting at the pointed snout. There are two separate dorsal fins. The lateralline is slightly arched anteriorly. The major distinguishing feature of the species is an adipose eyelid.

Nutritional Facts

Proximate Composition (per 100 g of Edible Portion)

Energy (Kcal)	Moisture (%)	Protein (g)	Fat (g)	Carbohydrate (g)	Ash (g)
93	76.8	20.6	1.1	0.2	1.3

Minerals (mg)

Ca	P	Fe	Na	K
0	101	1.1	52	300

Vitamins (mg)

Thiamine	Riboflavin	Niacin	Ascorbic Acid
0.08	0.10	3.2	0.09

Source: Data from Siong et al. (1987).

Chorinemus lysan (Forsskål, 1775)

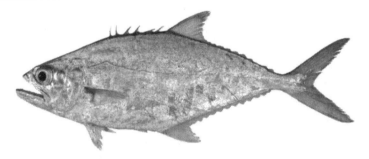

Order: Perciformes

Family: Carangidae

Common name: Doublespotted queen fish

Distribution: Indo-Pacific

Habitat: Marine; brackish; reef associated; depth range 0–100 m

Description: This species has a maximum length of 110 cm and weight of 11 kg. Mainly solitary, sometimes this species forms small loose groups. Adults feed on small fishes and crustaceans, while juveniles feed on scales and epidermal tissues torn from other schooling fishes.

Nutritional Facts

Proximate Composition (per 100 g of Edible Portion)

Energy (Kcal)	Moisture (%)	Protein (g)	Fat (g)	Carbohydrate (g)	Ash (g)
94	76.2	20.5	1.1	0.6	1.6

Minerals (mg)

Ca	P	Fe	Na	K
146	319	0.7	81	331

Vitamins (mg)

Thiamine	Riboflavin	Niacin	Ascorbic Acid
0.18	0.13	3.8	1.2

Source: Data from Siong et al. (1987).

Caranx djeddaba (Forsskål, 1775)

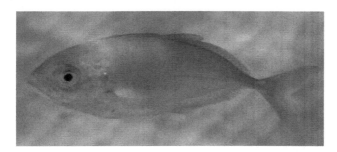

Order: Perciformes

Family: Carangidae

Common name: Yellowtail scad

Distribution: Eastern part of Mediterranean; Red Sea to the central Pacific Ocean

Habitat: Coastal waters; a schooling species

Description: This species has a well-developed adipose eyelid, which is more extensive posteriorly. Teeth in both jaws consist of a single row of continuous, small, comb-like teeth. The spinous dorsal fin is moderately high. Curved lateral line has 31–36 scales and 0–3 scutes, and the straight lateral line has 41–48 scutes. Color of the body is grayish-green above, silvery to white below. Normal size is 20 cm.

Nutritional Facts

Proximate Composition (per 100 g of Edible Portion)

Energy (Kcal)	Moisture (%)	Protein (g)	Fat (g)	Carbohydrate (g)	Ash (g)
103	75.4	21.5	1.9	0	1.4

Minerals (mg)

Ca	P	Fe	Na	K
80	252	0.8	55	303

Vitamins (mg)

Thiamine	Riboflavin	Niacin	Ascorbic Acid
0.11	0.26	4.1	0

Source: Data from Siong et al. (1987).

159

Decapterus russelli (Rüppell, 1830)

Order: Perciformes

Family: Carangidae

Common name: Indian shad

Distribution: Indo-West Pacific

Habitat: Benthopelagic; depth range 40–275 m

Description: This species has a maximum length of 45 cm and weight of 110 g. Lateral line is curved below soft dorsal and has 30–44 strong scutes. Color of the body is bluish green above and silvery below. Caudal fin is hyaline to yellowish. Dorsal fins are hyaline basally and light dusky distally. Upper jaw has small teeth anteriorly. Soft dorsal and anal fins are relatively low, not falcate. Pectoral fin is subfalcate.

Nutritional Facts

Proximate Composition

Moisture (%)	Protein (%)	Fat (%)	Ash (%)
66–75.5	20.6–21.9	1.2–10.7	1.4–1.6

Source: Data from Bykov (1983).

Fatty Acids of White Muscle

Saturated fatty acid (SFA) (%)	32.66%
Palmitic acid*	18.36%
Stearic acid*	12.21%
Monounsaturated fatty acids (MUFA) (%)	16.07%
Oleic acid*	10.61%
Palmotoleic acid*	3.91%
Polyunsaturated fatty acids (PUFA) (%)	39.38%
Eicosapentaenoic acid (EPA)*	8.83%
Docosahexaenoic acid (DHA)*	19.76%

* Major components of SFA, MUFA and PUFA.
Source: Data from Chrishanthi and Attygalle (2011).

Seriola quinqueradiata Temminck Schlegel, 1845

Order: Perciformes

Family: Carangidae

Common name: Japanese amberjack, yellowtail

Distribution: Northwest Pacific

Habitat: Demersal; oceanodromous

Description: This species has a maximum length of 150 cm and a weight of 40 kg. Dorsal soft rays and anal soft rays number 36 and 22, respectively. Scutes are absent. The dorsoposterior corner of the upper jaw is angular, and the pectoral and pelvic fins are almost equal in length.

Nutritional Facts

Proximate Composition

Protein (%)	Fat (%)	Carbohydrate (%)	Cholesterol (%)
42.8	27.1	0.03	24
(21.4 g)	(17.6 g)	(0.1 g)	(72 mg)

Seriolella punctata (Forster, 1801)

Order: Perciformes

Family: Centrolophidae

Common name: Silver warehou

Distribution: Eastern Indian Ocean; Southwest Pacific and Southeast Pacific Ocean

Habitat: Marine; brackish; benthopelagic; oceanodromous; depth range 27–650 m

Description: This species has a maximum length of 66 cm. It forms feeding and spawning aggregations. It is a schooling species, usually aggregating close to the sea bed. Adults mainly eat planktonic tunicates.

Nutritional Facts

Proximate Composition

Moisture (%)	Ash (%)	Lipid (%)	Protein (%)	Energetic Value (kJ/g Wet Mass)
63.96	2.15	14.78	14.83	9.36

Values based on percentage of wet mass.
Source: Data from Eder and Lewis (2005).

Fatty Acids

Saturated fatty acids	29.9%
Monounsaturated	46%
Polyunsaturated	24.1%

Nemadactylus bergi (Norman, 1937)

Order: Perciformes

Family: Cheilodactylidae

Common name: Hawkfish

Distribution: Southeast Pacific

Habitat: Demersal

Description: This species has a maximum length of 40 cm and weight of 450 g. Dorsal soft rays and anal soft rays number 28 and 15, respectively. It feeds on benthic invertebrates.

Nutritional Facts

Proximate Composition

Moisture (%)	Ash (%)	Lipid (%)	Protein (%)	Energetic Value (kJ/g Wet Mass)
62.83	4.98	13.25	15.13	8.83

Values based on percentage of wet mass.
Source: Data from Eder and Lewis (2005).

Drepane punctata (Linnaeus, 1758)

Order: Perciformes

Family: Drepaneidae

Common name: Spotted sicklefish

Distribution: Indo-West Pacific

Habitat: Marine; brackish; reef associated; amphidromous; depth range 10–49 m

Description: This species grows to a maximum length of 50 cm. Color of the body is generally silvery with a greenish tinge above. Pectoral fins are long and pointed. Four to eleven vertical gray spots are seen on the upper half of the sides, and generally eight dorsal spines are present.

Nutritional Facts

Proximate Composition (per 100 g of Edible Portion)

Energy (Kcal)	Moisture (%)	Protein (g)	Fat (g)	Carbohydrate (g)	Ash (g)
92	76.3	19.8	0.5	2.1	1.3

Minerals (mg)

Ca	P	Fe	Na	K
33	198	0.5	82	299

Vitamins (mg)

Thiamine	Riboflavin	Niacin	Ascorbic Acid
0.01	0.11	3.5	2.0

Source: Data from Siong et al. (1987).

Thalassoma fuscum (Lacepède, 1801) (= *Thalassoma trilobatum*)

Order: Perciformes

Family: Labridae

Common name: Christmas wrasse, Ladder wrasse

Distribution: Indo-Pacific

Habitat: Reef-associated; depth range 0–10 m

Description: This species grows to a maximum length of 30 cm. Head of male is plain brown to orange or shaded with blue. Female has a more spotted and shorter head. It lacks the 'V' on the snout. A diagonal dark red line is seen below and in front of the eye.

Nutritional Facts

Proximate Composition

Moisture (%)	Protein (%)	Fat (%)	Ash (%)	Carbohydrate (%)	Energy Value (Kcal/100 g)
77.28	17.47	1.52	1.66	0.38	90.28

Minerals (mg/100 g)

Ca	Fe	P
155.47	0.18	294.36

Source: Data from Palanikumar et al. (2014).

Cheilinus undulatus Rüppell, 1835

Order: Perciformes

Family: Labridae

Common name: Humphead wrasse

Distribution: Indo-Pacific

Habitat: Reef-associated; depth range 1–100 m

Description: This species grows to a maximum length of 230 cm and weight of 190 kg. Adults of this species develop thick lips and a prominent bulbous hump on the forehead. There are two black lines posteriorly from the eye.

Nutritional Facts

Proximate Composition (Dry Weight Basis)

Protein (%)	Carbohydrate (%)	Lipid (%)	Ash (%)
10.5	4.2	5.3	1.2

Vitamins (mg/100 g)

B_1	B_2	B_3	B_5	B_6	FA	A	K	D_3
0.44	0.48	0.61	0.19	0.56	0.92	0.33	0.55	0.62

FA: folic acid.

Source: Data from Dhaneesh et al. (2012).

Lates calcarifer (Bloch, 1790)

Order: Perciformes

Family: Latidae

Common name: Barramudi, giant sea-perch

Distribution: Indo-West Pacific

Habitat: Marine; freshwater; brackish; demersal; catadromous; depth range 10–40 m

Description: This species grows to a maximum length of 200 cm and weight of 60 kg. Body is elongate. Mouth is large and slightly oblique, and upper jaw is extending behind the eye. Caudal fin is rounded.

Nutritional Facts

Proximate Composition (per 100 g of Edible Portion)

Energy (Kcal)	Moisture (%)	Protein (g)	Fat (g)	Carbohydrate (g)	Ash (g)
83	78.1	19.5	0.1	1.0	1.3

Minerals (mg)

Ca	P	Fe	Na	K	Mg
28	220	0.4	76	416	808

Co	Cu	Mn	Zn
0.34	39.76	8.62	299.51

Source: Data from Nurnadia et al. (2013).

Vitamins (mg)

Thiamine	Riboflavin	Niacin	Ascorbic Acid
0.20	0.20	2.4	0

Source: Data from Siong et al. (1987).

Fatty Acids

EPA (µg/g)	DHA (µg/g)
55.4	109.4

EPA: eicosapentaenoic acid; DHA: docosa-hexaenoic acid.

Source: Data from Wan Rosli et al. (2012).

Leiognathus dussumieri (Valenciennes, 1835) (= *Karalla dussumieri*)

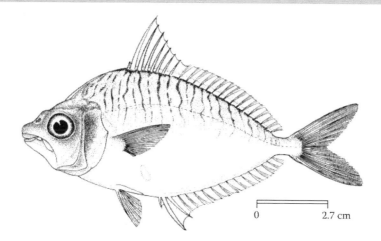

0   2.7 cm

Order: Perciformes

Family: Leiognathidae

Common name: Dussumier's ponyfish

Distribution: Indo-West Pacific

Habitat: Marine; brackish; demersal; depth range 10–40 m

Description: This species grows to a maximum length of 14 cm. It is found in coral sand bottoms of coastal waters, but also enters estuaries. It feeds on small crustaceans, polychaetes, bivalves, foraminiferans, gastropods, and nematodes and forms schools.

Gazza achlamys Jordan Starks, 1917

Order: Perciformes

Family: Leiognathidae

Common name: Smalltoothed ponyfish

Distribution: Indo-West Pacific

Habitat: Marine; brackish; reef associated

Description: This species grows to a maximum length of 17 cm. There are eight dorsal spines and 16 dorsal soft rays. Anal spines and anal soft rays number 3 and 14, respectively. Membrane of spinous dorsal fin is black at its distal portion. It feeds mainly on small fishes, crustaceans, and polychaetes and forms schools.

Nutritional Facts

Proximate Composition

Species	Moisture (%)	Protein (%)	Fat (%)	Ash (%)	Carbohydrate (%)	Energy Value (Kcal/100 g)
Leiognathus dussumieri	67.23	10.24	14.72	1.20	0.080	174.16
Gazza achlamys	77.34	19.30	0.79	2.40	0.07	93.91

Minerals (mg/100 g)

Species	Ca	Fe	P
Leiognathus dussumieri	90.41	2.66	190.27
Gazza achlamys	200.08	3.95	182.20

Source: Data from Palanikumar et al. (2014).

Leiognathus equulus (Forsskål, 1775)

Order: Perciformes

Family: Leiognathidae

Common name: Greater pony fish

Distribution: Indo-West Pacific

Habitat: Marine; freshwater; brackish; demersal; amphidromous; depth range 10–110 m

Description: This species grows to a maximum length of 28 cm. Adults are coastal inhabitants found on soft bottoms, usually within depths of 10–70 m. Juveniles are commonly found in mangrove estuaries and tidal creeks, sometimes entering the lower reaches of freshwater streams. Adults move in schools. This species feeds on polychaetes, small crustaceans, small fishes, and worms.

Nutritional Facts

Proximate Composition (per 100 g of Edible Portion)

Energy (Kcal)	Moisture (%)	Protein (g)	Fat (g)	Carbohydrate (g)	Ash (g)
77	79.8	19.3	0	0	1.4

Minerals (mg)

Ca	P	Fe	Na	K
54	213	0.6	99	227

Vitamins (mg)

Thiamine	Riboflavin	Niacin	Ascorbic Acid
0.01	0.07	2.8	0

Source: Data from Siong et al. (1987).

167

Lethrinus lentjan (Lacepède, 1802)

Order: Perciformes

Family: Lethrinidae

Common name: Red spotted emperor

Distribution: Indo-West Pacific

Habitat: Marine; brackish; reef associated; nonmigratory; depth range 10–90 m

Description: This species grows to a maximum length of 52 cm. Mouth is slightly protractile, and lips are thick and fleshy. Body is olive-green above, becoming paler below. The pectoral fin is white, yellow, or pinkish. The pelvic and anal fins are white to orange. The dorsal fin is white and orange and is mottled with a reddish margin. The caudal fin is mottled orange or reddish.

Nutritional Facts

Proximate Composition (mg/gm)

Protein	Fat	Carbohydrate
23.84	9.88	1.75

Source: Data from Mathana et al. (2012).

Lutjanus quinquelineatus (Bloch, 1790)

Order: Perciformes

Family: Lutjanidae

Common name: Five-lined snapper, blue-striped snapper

Distribution: Indo-West Pacific

Habitat: Reef associated; depth range 2–40 m

Description: This species grows to a maximum length of 38 cm. Dorsal profile of head is steeply sloped. Scale rows on back are rising obliquely above the lateral line. Body color is generally bright yellow, including fins, with a series of blue stripes on the side. A round black spot, about the size of the eye or larger, is seen below the anterior-most soft dorsal rays, touching the lateral line.

Lutjanus lutjanus (Bloch, 1790)

Order: Perciformes

Family: Lutjanidae

Common name: Bigeye snapper

Distribution: Indo-West Pacific

Habitat: Bathydemersal; depth range 0–96 m

Description: This species grows to a maximum length of 35 cm. Dorsal profile of head is gently sloped. Scale rows on back are rising obliquely above lateral line. Body color is generally silvery white, with a broad yellow stripe running along the side from the eye to the caudal fin base. A series of faint narrow yellow horizontal lines is on the lower half of the body. The fins are pale yellow to whitish.

Lutjanus decussatus (Cuvier, 1828)

Order: Perciformes

Family: Lutjanidae

Common name: Checkered snapper

Distribution: Indo-West Pacific

Habitat: Reef-associated; depth range 2–35 m

Description: This species grows to a maximum length of 35 cm. Dorsal profile of head is moderately sloped. Scale rows on back are rising obliquely above lateral line. Body color is generally pink to whitish, with a silvery sheen on breast and side of head. Trunk has a "checkerboard" pattern on upper half of sides. Five to six horizontal brown stripes are seen. Caudal fin base has a large black spot.

Nutritional Facts

Proximate Composition

Species	Moisture (%)	Protein (%)	Fat (%)	Ash (%)	Carbohydrate (%)	Energy (Kcal/100 g)
Lutjanus quinquelineatus	75.75	21.46	3.43	1.81	0.04	116.87
Lutjanus lutjanus	76.29	15.67	0.24	1.54	0.05	65.04
Lutjanus decussatus	73.52	21.13	2.06	0.98	0.06	106.98

Minerals (mg/100 g)

	Ca	Fe	P
Lutjanus quinquelineatus	721.10	18.69	214.66
Lutjanus lutjanus	1135.05	2.95	305.25
Lutjanus decussatus	1887.10	3.08	204.70

Source: Data from Palanikumar et al. (2014).

Lutjanus argentimaculatus (Forsskål, 1775)

Order: Perciformes

Family: Lutjanidae

Common name: Mangrove red snapper

Distribution: Indo-West Pacific

Habitat: Marine; freshwater; brackish; reef associated; oceanodromous; depth range 1–120 m

Description: This species grows to a maximum length of 150 cm and weight of 8.7 kg. Scale rows on the back are more or less parallel to the lateral line. Body is generally greenish-brown on back, grading to reddish on the sides and ventral parts. Dorsal fin is greenish-brown, and ventral fin is white or greenish-gray. Sides are reddish.

Nutritional Facts

Proximate Composition (per 100 g of Edible Portion)

Energy (Kcal)	Moisture (%)	Protein (g)	Fat (g)	Carbohydrate (g)	Ash (g)
102	75.9	21.0	1.9	0.1	1.1

Minerals (mg)

Ca	P	Fe	Na	K
21	189	0.5	80	327

Vitamins (mg)

Thiamine	Riboflavin	Niacin	Ascorbic Acid
0.03	0.08	3.1	0.08

Source: Data from Siong et al. (1987).

Lutjanus gibbus (Forsskål, 1775)

Order: Perciformes

Family: Lutjanidae

Common name: Humpback red snapper

Distribution: Indo-Pacific.

Habitat: Reef-associated; depth range 1–150 m

Description: This species grows to a maximum length of 50 cm. Adults mainly inhabit coral reefs, sometimes forming large aggregations, which are mostly stationary during the day. It feeds on fishes and a variety of invertebrates including shrimps, crabs, lobsters, stomatopods, cephalopods, echinoderms, and ophiuroids.

Nutritional Facts

Proximate Composition (Dry Weight Basis)

Protein (%)	Carbohydrate (%)	Lipid (%)	Ash (%)
10.6	3.0	4.8	1.3

171

Vitamins (mg/100 g)

B₁	B₂	B₃	B₅	B₆	FA	A	K	D₃
0.48	0.56	0.52	0.44	0.58	0.84	0.33	0.41	0.51

FA: folic acid.

Source: Data from Dhaneesh et al. (2012).

Lutjanus russelli (Bleeker, 1849)

Order: Perciformes

Family: Lutjanidae

Common name: Russell's snapper

Distribution: Western Pacific: Indian Ocean

Habitat: Marine; brackish; reef-associated; depth range 3–80 m

Description: This species grows to a maximum length of 50 cm. Dorsal profile of head is steeply to moderately sloped. Scale rows on back are rising obliquely above the lateral line. Body is generally whitish or pink with silvery sheen, frequently brownish on the upper part of the head and back. A black spot, which is sometimes faint, is on the lateral line below the anterior portions of the soft dorsal fin.

Nutritional Facts

Proximate Composition (per 100 g of Edible Portion)

Energy (Kcal)	Moisture (%)	Protein (g)	Fat (g)	Carbohydrate (g)	Ash (g)
8	77.0	20.1	0.3	1.3	1.3

Minerals (mg)

Ca	P	Fe	Na	K
80	250	0.5	85	321

Vitamins (mg)

Thiamine	Riboflavin	Niacin	Ascorbic Acid
0.04	0.10	2.2	0

Source: Data from Siong et al. (1987).

Pristipomoides typus Bleeker, 1852

Order: Perciformes

Family: Lutjanidae

Common name: Sharptooth bass

Distribution: Eastern Indian Ocean and Western Pacific

Habitat: Demersal; depth range 40–120 m

Description: This species grows to a maximum length of 70 cm and weight of 2.2 kg. Pectoral fins are long, reaching level of anus. Scale rows on back are parallel to lateral line. Overall color of body is rosy red. Top of the head has longitudinal vermiculated lines and spots of brownish-yellow. Dorsal fin has wavy yellow lines.

Nutritional Facts

Proximate Composition (per 100 g of Edible Portion)

Energy (Kcal)	Moisture (%)	Protein (g)	Fat (g)	Carbohydrate (g)	Ash (g)
93	77.2	20.2	1.3	0	1.4

Minerals (mg)

Ca	P	Fe	Na	K
45	214	0.3	53	217

Vitamins (mg)

Thiamine	Riboflavin	Niacin	Ascorbic Acid
0.04	0.04	3.9	0

Source: Data from Siong et al. (1987).

Lutjanus johnii (Bloch, 1792)

173

Order: Perciformes

Family: Lutjanidae

Common name: Golden snapper

Distribution: Indo-West Pacific

Habitat: Marine; brackish; reef-associated; oceanodromous

Description: This species grows to a maximum length of 97 cm and weight of 10.5 kg. Dorsal profile of head is steeply sloped. Center of each scale often has a reddish-brown spot. Body is generally yellow with a bronze to silvery sheen, shading to silvery white on belly and underside of the head. A large black blotch is present above the lateral line.

Nutritional Facts

Proximate Composition

Energy (kJ/g)	Moisture (%)	Protein (%)	Fat (%)	Ash (%)
5.1	80.2	19.4	1.3	1.1

Source: Data from Nurnadia et al. (2011).

Minerals (mg/100 g)

Na	K	Ca	Mg	Co	Cu	Fe	Mn	Zn
27.3	12.8	21.4	66.0	0.7	68.6	351.7	7.5	164

Source: Data from Nurnadia et al. (2011, 2013).

Lutjanus malabaricus (Bloch & Schneider, 1801)

Order: Perciformes

Family: Lutjanidae

Common name: Malabar blood snapper

Distribution: Indo-West Pacific

Habitat: Reef associated; depth range 12–100 m

Description: This species grows to a maximum length of 100 cm and weight of 7.9 kg. Adults inhabit both coastal and offshore reefs. They tend to be associated with sponge and gorgonian-dominated habitats and hard mud areas. They frequently form mixed shoals with *L. erythropterus*. They feed mainly on fishes, with small amounts of benthic crustaceans.

Nutritional Facts

Proximate Composition (per 100 g of Edible Portion)

Energy (Kcal)	Moisture (%)	Protein (g)	Fat (g)	Carbohydrate (g)	Ash (g)
104	75.5	20.8	2.3	0.1	1.3

Vitamins (mg)

Thiamine	Riboflavin	Niacin	Ascorbic Acid
0.08	0.10	3.4	0

Minerals (mg)

Ca	P	Fe	Na	K	Mg
26	216	0.3	81	359	898

Co	Cu	Mn	Zn
0.28	197.06	16.58	259.54

Source: Data from Siong et al. (1987).

Source: Data from Nurnadia et al. (2013).

Lutjanus bohar (Forsskål, 1775)

Order: Perciformes

Family: Lutjanidae

Common name: Two-spot red snapper

Distribution: Indo-Pacific

Habitat: Reef-associated; depth range 4–180 m

Description: This species grows to a maximum length of 90 cm and weight of 12.5 kg. Snout is somewhat pointed. Dorsal profile of head is rounded. Scale rows on back are rising obliquely above the lateral line. Young and some adults have two silvery-white spots on back. Large adults are mostly plain red.

Nutritional Facts

Proximate Composition (Dry Weight Basis)

Protein (%)	Carbohydrate (%)	Lipid (%)	Ash (%)
11.7	5.9	4.5	1.7

Vitamins (mg/100 g)

B₁	B₂	B₃	B₅	B₆	FA	A	K	D₃
0.48	0.37	0.87	0.18	0.91	0.55	0.49	0.53	0.42

FA: folic acid.
Source: Data from Dhaneesh et al. (2012).

175

Monodactylus argenteus (Linnaeus, 1758)

Order: Perciformes

Family: Monodactylidae

Common name: Moon fish

Distribution: Indo-West Pacific

Habitat: Marine; freshwater; brackish; pelagic-neritic

Description: This species grows to a maximum length of 27 cm. Adults are bright silver with yellow and dusky dorsal fin tip. Fins are yellow except pectoral translucent. Eyes are large, and mouth is small.

Nutritional Facts

Proximate Composition

Moisture (%)	Ash (%)	Lipid (%)	Protein (%)	Energetic Value (kJ/g Wet Mass)
74.6	1.2	6.9	19.6	7.4

Source: Data from Nurnadia et al. (2011).

Minerals (mg/100 g)

Ca	Na	K	Mg	Co	Cu	Fe	Mn	Zn
0.04	95.13	331.05	12.71	303.41	29.95	22.42	12.76	670.14

Source: Data from Nurnadia et al. (2013).

176

Morone saxatilis (Walbaum, 1792)

Order: Perciformes

Family: Moronidae

Common name: Striped bass

Distribution: Coastal waters; anadromous

Habitat: Atlantic coastline of North America from the St. Lawrence River into the Gulf of Mexico to approximately Louisiana

Description: It has a streamlined, silvery body marked with longitudinal dark stripes running from behind the gills to the base of the tail. The maximum scientifically recorded weight is 57 kg, and its maximum length is 1.8 m.

Nutritional Facts

Proximate Composition (g/100g WW)

Moisture	Protein	Oil	Ash	Energy (Kcal)
79.22	17.73	2.33	1.04	97

Minerals (mg/100 g WW)

Ca	Fe	Mg	P	K	Na	Zn	Cu	Mn
15.0	0.84	4.0	198	256	69	0.40	0.03	0.02

Source: Data from Stansby (1987).

Fatty Acids (g/100 g WW)

SFA	MFA	PUFA
0.51	0.66	0.78

Vitamins (per 100 g WW)

Thiamine	0.10 mg
Riboflavin	0.03 mg
Niacin	2.10 mg
Pantothenic acid	0.75 mg
B_1	0.30 mg
Folate	9.0 µg
B_{12}	3.82 µg
A	90 IU

Amino acids (per 100 g WW)

Tryptophan	0.199 g
Threonine	0.777 g
Isoleucine	0.817 g
Leucine	1.441 g
Lysine	1.628 g
Methionine	0.525 g
Cystine	0.190 g
Phenylalanine	0.692 g
Tyrosine	0.599 g
Valine	0.914 g
Arginine	1.061 g
Histidine	0.522 g
Alanine	1.072 g
Aspartic acid	1.816 g
Glutamic acid	2.647 g
Glycine	0.851 g
Proline	0.627 g
Serine	0.723 g

Source: Data from the *Titi Tudoranceea Bulletin* (1991–2014).

Dicentrarchus labrax (Linnaeus, 1758)

Order: Perciformes

Family: Moronidae

Common name: European sea bass

Distribution: Eastern Atlantic; Mediterranean and Black Sea.

Habitat: Marine; freshwater; brackish; demersal; oceanodromous; depth range 10–100 m

Description: This species grows to a maximum length of 103 cm and weight of 12 kg. Mouth is moderately protractile. Young have some dark spots on upper part of body. Head has cycloid scales above. Vomerine teeth are present only anteriorly, in a crescentic band.

Nutritional Facts

Proximate Composition

Moisture (%)	Ash (%)	Lipid (%)	Protein (%)
75.5	1.5	1.4	19.2

Fatty Acids

Saturated	31.1%
Monounsaturated	23.2%
Polyunsaturated	44%
EPA + DHA	31.2%

Source: Data from Nasopoulou et al. (2011).

Parupeneus bifasciatus (Lacepède, 1801)

Order: Perciformes

Family: Mullidae

Common name: Yellow spot goat fish

Distribution: Indian Ocean

Habitat: Reef-associated; depth range 1–80 m

Description: This species grows to a maximum length of 35 cm. The total number of dorsal spines, dorsal soft rays, anal spines, and anal soft rays is 8, 9, 1, and 7, respectively. This occasionally schooling species inhabits lagoon and seaward reefs, and juveniles inhabit reef flats. Adults tend to occur around rocky or coralline areas of high vertical relief. This species is known to occur in three color forms. For example, the Indo-Australian form is pale with a dark blotch around the eye and two dark bars on the body; the Pacific form ranges from brown to red with two distinct bars; and the Indian Ocean form has two wedge-shaped bars.

Nutritional Facts

Proximate Composition (Dry Weight Basis)

Protein (%)	Carbohydrate (%)	Lipid (%)	Ash (%)
10.6	6.1	5.6	1.6

Vitamins (mg/100 g)

B_1	B_2	B_3	B_5	B_6	FA	A	K	D_3
0.54	0.44	0.19	0.53	0.67	0.63	0.62	0.41	0.51

FA: folic acid.

Source: Data from Dhaneesh et al. (2012).

Nemipterus bleekeri (Day, 1875) (= *Nemipterus bipunctatus*)

Order: Perciformes

Family: Nemipteridae

Common name: Delagoa threadfin bream, threadfin bream

Distribution: Indian Ocean

Habitat: Demersal; nonmigratory; depth range 18–100 m

Description: This species grows to a maximum length of 30 cm. Pectoral fins are moderately long, reaching to or just behind anus. Pelvic fins are long, reaching to or just beyond the level of origin of the anal fin. Scale rows on body below the lateral line are upward-curved anteriorly. Axillary scale is present. Upper body pinkish and silvery below. Snout has two indistinct stripes.

179

Nemipterus japonicus (Bloch, 1791)

Order: Perciformes

Family: Nemipteridae

Common name: Japanese threadfin bream

Distribution: Indo-Pacific

Habitat: Demersal; nonmigratory; depth range 5–80 m

Description: This species grows to a maximum length of 32 cm and weight of 600 g. Eleven to twelve pale golden-yellow stripes are seen along the body from behind the head to the base of the caudal fin. A prominent red-suffused yellow blotch is present below the origin of the lateral line. Pectoral fins are very long, reaching to or just beyond the level of origin of the anal fin. Caudal fin is moderately forked. Upper body is pinkish and silvery below.

Scolopsis bimaculatus Rüppell, 1828

Order: Perciformes

Family: Nemipteridae

Common name: Thumbprint threadfin bream

Distribution: Indian Ocean

Habitat: Reef associated

Description: This species grows to a maximum length of 31 cm. Head scales

180

reach forward to level of posterior nostrils. Pelvic fins are long, reaching to between level of anus and origin of anal fin. Axillary scale is present. Upper body is gray, and whitish below. A blue stripe joins the eyes.

Nutritional Facts

Minerals (mg/100 g)

	Ca	Fe	P
Nemipterus bleekeri	64.24	4.11	203.43
Nemipterus japonicus	0798.63	0.16	255.16
Scolopsis bimaculatus	1175.00	0.42	266.54

Source: Data from Palanikumar et al. (2014).

Proximate Composition (per 100 g of Edible Portion)

Energy (Kcal)	Moisture (%)	Protein (g)	Fat (g)	Carbohydrate (g)	Ash (g)
83	78.7	18.8	0.9	0	1.7

Vitamins (mg)

Thiamine	Riboflavin	Niacin	Ascorbic Acid
0.03	0.13	2.1	0

Source: Data from Siong et al. (1987).

Minerals (mg)

Ca	P	Fe	Na	K	Co	Cu	Mn	Zn
10	92.81	13.38	147.87	53	200	0.8	246	279

Source: Data from Siong et al. (1987) and Nurnadia et al. (2013).

Fatty Acids

EPA (µg/g)	DHA (µg/g)
32.6	115.5

EPA: eicosapentaenoic acid; DHA: docosahexaenoic acid

Source: Data from Wan Rosli et al. (2012).

Dissostichus eleginoides Smitt, 1898

Order: Perciformes

Family: Nototheniidae

Common name: Chilean sea bass

Distribution: Southeast Pacific and Southwest Atlantic Oceans; Southwest Pacific Ocean

Habitat: Pelagic-oceanic; oceanodromous; depth range 50–3850 m

Description: This species grows to a maximum length of 215 cm and weight of 9.6 kg. The maximum number of dorsal spines, dorsal soft rays, and anal soft rays is 10, 30, and 30, respectively. The number of lower and upper lateral line scales is 104 and 77, respectively. Adults migrate to deeper habitats at depths greater than 1000 m.

Nutritional Facts

Proximate Composition (Values Based on Percentage of Wet Mass)

Moisture (%)	Ash (%)	Lipid (%)	Protein (%)	Energetic Value (kJ/g Wet Mass)
61.42	2.50	17.06	15.64	10.46

Source: Data from Eder and Lewis (2005).

Patagonotothen ramsayi (Regan, 1913)

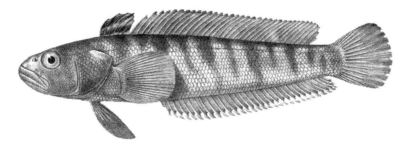

Order: Perciformes

Family: Nototheniidae

Common name: Longtail southern cod

Distribution: Southwest Atlantic: restricted to the Patagonian shelf.

Habitat: Benthopelagic; depth range 50–500 m

Description: This species grows to a maximum length of 44 cm. The maximum number of dorsal spines, dorsal soft rays, and anal soft rays is 7, 35, and 34, respectively. The number of lower and upper lateral line scales is 54 and 17, respectively. Caudal fin is rounded.

Nutritional Facts

Proximate Composition (Values Based on Percentage of Wet Mass)

Moisture (%)	Ash (%)	Lipid (%)	Protein (%)	Energetic Value (kJ/g Wet Mass)
73.06	3.10	7.36	17.07	6.88

Source: Data from Eder and Lewis (2005).

Fatty Acids

Docosahexaenoic acid	30–42%
Eicosapentaenoic acid	12–18%

Source: Data from González et al. (2007).

Perca fluviatilis **Linnaeus, 1758**

Order: Perciformes

Family: Percidae

Common name: Perch

Distribution: Eurasia

Habitat: Brackish water and fresh water; demersal

Description: This species grows to a maximum length of 60 cm and a weight of 4.8 kg. It is distinguished by its pelvic and anal fins which are yellow to red. Posterior part of first dorsal fin has a dark blotch. Flank has 5–8 bold dark bars, usually Y-shaped. The two dorsal fins are clearly separated from each other. Body is greenish-yellow. First dorsal fin is gray with black spot at the tip. Second dorsal is greenish-yellow, and pectorals are yellow. Other fins are red. Caudal fin is emarginate.

Nutritional Facts

Proximate Composition (per 100 g)

Calories	Protein (g)	Fat (g)
105	19	3

Fatty Acids (% of Total Fatty Acids)

PUFA	DHA	EPA
43.0	16.7	7.3

PUFA: polyunsaturated fatty acids; DHA: docosahexaenoic acid; EPA: eicosapentaenoic acid.
Source: Data from Hossain (2011).

Pseudopercis semifasciata **(Cuvier, 1829)**

Order: Perciformes

Family: Pinguipedidae

Common name: Sea salmon, Argentinian sandperch

Distribution: Southwest Atlantic

Habitat: Demersal; rocky and sandy bottoms in coastal waters

Description: This species grows to a maximum length of 100 cm and a weight of 10 kg. Body is pale brown with several dark brown vertical and horizontal stripes composed of dark brown blotches. A row of dark brown blotches is seen on middle of dorsal fin membrane. An obvious black blotch is present on the base of the upper lobe of the caudal fin.

Nutritional Facts

Proximate Composition (Values Based on Percentage of Wet Mass)

Moisture (%)	Ash (%)	Lipid (%)	Protein (%)	Energetic Value (kJ/g Wet Mass)
73.94	3.70	3.84	14.56	4.97

Source: Data from Eder and Lewis (2005).

Plectorhinchus pictus (Tortonese, 1936)

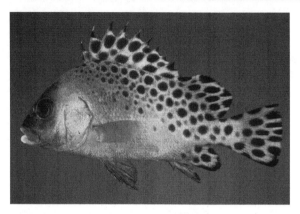

Order: Perciformes

Family: Haemulidae

Common name: Painted sweetlip

Distribution: Indo-West Pacific

Habitat: Reef-associated; depth range 20–200 m

Description: This species grows to a maximum length of 83 cm and a weight of 6.9 kg. These fish have big, fleshy lips. They are usually seen in clusters in nooks and crannies or under over-hangs. Sweetlips' coloring and pattern-ing change throughout their lives.

Nutritional Facts

Proximate Composition (per 100 g of Edible Portion)

Energy (Kcal)	Moisture (%)	Protein (g)	Fat (g)	Carbohydrate (g)	Ash (g)
83	78.2	19.8	0.2	0.5	1.3

Minerals (mg)

Ca	P	Fe	Na	K
39	211	0.4	84	365

Vitamins (mg)

Thiamine	Riboflavin	Niacin	Ascorbic Acid
0.01	0.09	3.1	0

Source: Data from Siong et al. (1987).

Pomadasys hasta (Bloch, 1790)

Order: Perciformes

Family: Haemulidae

Common name: Silver grunter

Distribution: Indo-West Pacific

Habitat: Marine; freshwater; brackish; demersal; depth range 15–115 m

Description: This species grows to a maximum length of 70 cm. Body is ovate, and head profile is almost straight. Mouth is small, and lips are not thickened. Color is generally silver-mauve to fawn above and white below. Large specimens are plain or have scattered charcoal scale spots on back and upper sides. Snout is dark brown, and the upper operculum is purplish.

Nutritional Facts

Proximate Composition (per 100 g of Edible Portion)

Energy (Kcal)	Moisture (%)	Protein (g)	Fat (g)	Carbohydrate (g)	Ash (g)
84	78.3	19.5	0.4	0.5	1.3

Minerals (mg)

Ca	P	Fe	Na	K
23	219	0.7	58	460

Vitamins (mg)

Thiamine	Riboflavin	Niacin	Ascorbic Acid
0.01	0.10	3.0	0

Source: Data from Siong et al. (1987).

Eleutheronema tetradactylum (Shaw, 1804)

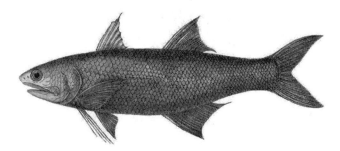

Order: Perciformes

Family: Polynemidae

Common name: Four-finger threadfin

Distribution: Indo-West Pacific

Habitat: Marine; freshwater; brackish; pelagic-neritic; amphidromous; depth range 0–23 m

Description: This species grows to a maximum length of 200 cm and a weight of 145 kg. Four pectoral filaments are seen. Fin membranes are vivid yellow in life, except in large specimens. Vomer has deciduous tooth plates on both sides, except in juveniles. Posterior part of maxilla is deep. Short tooth plate extends onto lateral surface of lower jaw.

Nutritional Facts

Proximate Composition (per 100 g of Edible Portion)

Energy (Kcal)	Moisture (%)	Protein (g)	Fat (g)	Carbohydrate (g)	Ash (g)
117	74.2	21.7	3.4	0	1.3

Vitamins (mg)

Thiamine	Riboflavin	Niacin	Ascorbic Acid
0.17	0.07	4.0	1.4

Source: Data from Siong et al. (1987).

Minerals (mg)

Ca	P	Fe	Na	K	Mg	Co	Cu	Mn	Zn
0.27	92.86	8.00	244.26	16	227	0.6	32	254	680

Source: Data from Siong et al. (1987) and Nurnadia et al. (2013).

Fatty Acids

Fat (g/100 g)	2.24
Cholesterol (mg/100 g)	45.9
SFA	8.14 %
MUFA	1.98 %
PUFA (ω6)	19.78 %
PUFA (ω3)	29.67 %
Other PUFA	40.00 %
Total fatty acids	89.5

SFA: saturated fatty acid; MUFA: monounsaturated fatty acids; PUFA: polyunsaturated fatty acids.
Source: Data from Osman et al. (2001).

Pomatomus saltatrix (Linnaeus, 1766)

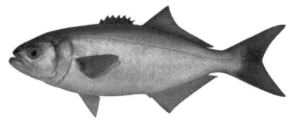

Order: Perciformes

Family: Pomatomidae

Common name: Bluefish

Distribution: Circumglobal: In tropical to subtropical waters; except the eastern Pacific

Habitat: Marine; brackish; pelagic-oceanic; oceanodromous; depth range 0–200 m

Description: This species grows to a maximum length of 130 cm and a weight of 14.4 kg. Jaw teeth are prominent, sharp, compressed, and in a single series. Of the two dorsal fins, the first dorsal is short and low, with seven or eight feeble spines connected by a membrane. Back is greenish, and sides and belly are silvery.

Nutritional Facts (per 100 g)

Water	70.860 g
Energy	124.000 Kcal
Energy	519.000 kJ
Protein	20.040 g
Total lipid (fat)	4.240 g
Ash	1.040 g
Calcium, Ca	7.000 mg
Iron, Fe	0.480 mg
Magnesium, Mg	33.000 mg
Phosphorus, P	227.000 mg
Potassium, K	372.000 mg
Sodium, Na	60.000 mg
Zinc, Zn	0.810 mg
Copper, Cu	0.053 mg
Manganese, Mn	0.021 mg
Selenium, Se	36.500 µg
Thiamin	0.058 mg
Riboflavin	0.080 mg
Niacin	5.950 mg
Pantothenic acid	0.828 mg
Vitamin B_6	0.402 mg
Folate, total	2.000 µg
Folate	2.000 µg_DFE
Vitamin B_{12}	5.390 µg
Vitamin A	398.000 IU
Vitamin A	120.000 µg_RAE
Retinol	120.000 µg
Fatty acids, total saturated	0.915 g
Fatty acids, total monounsaturated	1.793 g
Fatty acids, total polyunsaturated	1.060 g
Cholesterol	59.000 mg

RAE: retinol activity equivalent; DFE: dietary folate equivalent.
Source: Data from *Titi Tudorancea Bulletin* (2014).

Proximate Composition

Moisture (%)	Protein (%)	Fat (%)	Ash (%)
72	20	7	1.1

Source: Data from Stansby (1987).

Sciaenops ocellatus (Linnaeus, 1766)

Order: Perciformes

Family: Sciaenidae

Common name: Red drum

Distribution: Western Atlantic

Habitat: Marine; brackish; demersal; oceanodromous;

Description: This species grows to a maximum length of 155 cm and weight of 45 kg. Red drum has a dark red color on the back, which fades into white on the belly. The most distinguishing mark on the red drum is one large black spot on the upper part of the tail base.

Nutritional Facts

Proximate Composition

Protein (%)	Moisture (%)	Ash (%)	Fat (%)
17.8	77.4	1.48	0.62

Fatty Acids (% of Total Fatty Acids)

PUFA	DHA	EPA
37.3	21.1	3.6

PUFA: polyunsaturated fatty acids; DHA: docosahexaenoic acid; EPA: eicosapentaenoic acid.
Source: Data from Hossain (2011).

Atractoscion nobilis (Ayres, 1860) (= *Cynoscion nobilis*)

Order: Perciformes

Family: Sciaenidae

Common name: White weakfish

Distribution: Eastern Pacific

Habitat: Demersal; depth range 0–122 m

Description: This species grows to a maximum length of 166 cm and a weight of 41 kg. The total number of dorsal spines, dorsal soft rays, anal spines, and anal soft rays is 11, 23, 2, and 9, respectively. Pelvic fins have fleshy appendage at base. Young are seen in bays and along sandy beaches.

Cynoscion nebulosus (Cuvier, 1830)

Order: Perciformes

Family: Sciaenidae

Common name: Spotted seatrout

Distribution: Western Atlantic; Northeastern Atlantic

Habitat: Marine; brackish; demersal;

Description: This species grows to a maximum length of 100 cm and weight of 7.9 kg. Body is silvery, dark gray on back, with bluish reflections and numerous round black spots irregularly scattered on the upper half, extending to the dorsal and caudal fins. Spinous dorsal fin is dusky, and other fins are pale to yellowish. Mouth is large and oblique, and lower jaw is projecting. Snout has two marginal pores.

Nutritional Facts (per 100 g)

Total fat	4 g
Cholesterol	83 mg
Sodium	58 mg

Source: Data from Fatsecret (2014).

Proximate Composition

Moisture (%)	Protein (%)	Oil (%)	Ash (%)
77	19	2.6	1.1

Source: Data from Stansby (1987).

Cynoscion regalis (Bloch & Schneider, 1801)

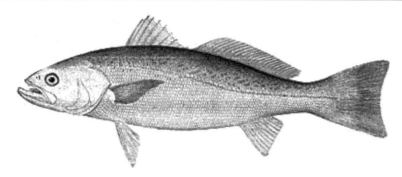

Order: Perciformes

Family: Sciaenidae

Common name: Squeteague, weakfish

Distribution: Western Atlantic

Habitat: Marine; brackish; demersal; oceanodromous; depth range 10–26 m

Description: This species grows to a maximum length of 98 cm and weight of 8.9 kg. Body is greenish-gray above and silvery below. Back has small spots forming undulating dotted lines. Pelvic fins and anal fin are yellowish and other fins are pale, sometimes with a yellowish tinge. Mouth is large and oblique, and lower jaw is projecting. Soft portion of dorsal fin is covered with small scales up to half of fin height.

Nutritional Facts (per 3 oz.)

Protein	18 g
Carbohydrate	0 g
Total fat	4 g
Saturated fat	1.1 g
Omega-3 fatty acids	0.4 g
Cholesterol	90 mg
Sodium	63 mg

Source: Data from USDA (1987).

Proximate Composition

Moisture (%)	Protein (%)	Oil (%)	Ash (%)
78	19	1.7	1.1

Source: Data from Stansby (1987).

Leiostomus xanthurus Lacepède, 1802

Order: Perciformes

Family: Sciaenidae

Common name: Spot croaker

Distribution: Western Atlantic

Habitat: Marine; brackish; demersal; oceanodromous;

Description: This species grows to a maximum length of 36 cm and weight of 450 g. It is a moderately deep-bodied, compressed fish with an elevated back. Body color is typically bluish-gray dorsally, fading to golden yellow or yellow-tan ventrally. A set of 12–15 dark streaks runs obliquely from the dorsal surface down the sides to about midbody. Fins are typically pale yellow in color. The head is short, with a small, inferior mouth. The dorsal fin is continuous. A large black spot is set above the upper edge of the gill cover.

Nutritional Facts

Proximate Composition

Moisture (%)	Protein (%)	Oil (%)	Ash (%)
75	18	6.0	1.1

Source: Data from Stansby (1987).

Johnius dussumieri (Cuvier, 1830) (= *Sciaena dussumieri*)

Order: Perciformes

Family: Sciaenidae

Common name: Sin croaker

Distribution: Indian Ocean: Pakistan to the Andaman Islands

Habitat: Marine; brackish; demersal; oceanodromous

Description: It has maximum and common lengths of 40 cm and 14 cm, respectively. It is found in coastal waters and enters estuaries. It feeds on invertebrates and small fishes.

Nutritional Facts

Proximate Composition (per 100 g of Edible Portion)

Energy (Kcal)	Moisture (%)	Protein (g)	Fat (g)	Carbohydrate (g)	Ash (g)
93	77.3	19.4	1.5	0.4	1.4

Minerals (mg)

Ca	P	Fe	Na	K
42	191	0.4	129	340

Vitamins (mg)

Thiamine	Riboflavin	Niacin	Ascorbic Acid
0.10	0.16	2.0	0

Source: Data from Siong et al. (1987).

Proximate Composition (Dryfish Values)

Moisture (%)	Protein (%)	Carbohydrate (%)	Oil (%)	Ash (%)
22.7	57.6	3.7	5.4	6.0

Source: Data from Siddique et al. (2012).

Nibea soldado (Lacepède, 1802) (= *Johnius (Pseudosciaena) soldado*)

191

Order: Perciformes

Family: Sciaenidae

Common name: Soldier croaker, silver jewfish

Distribution: Indo-West Pacific

Habitat: Marine; freshwater; brackish; demersal; amphidromous

Description: It has maximum and common lengths of 60 cm and 40 cm, respectively. Snout is rounded. Teeth are differentiated into large and small in both jaws. Second anal spine is long and stiff. Swim bladder is carrot shaped, with 18–22 pairs of arborescent appendages along its sides.

Nutritional Facts

Proximate Composition (per 100 g of Edible Portion)

Energy (Kcal)	Moisture (%)	Protein (g)	Fat (g)	Carbohydrate (g)	Ash (g)
97	76.8	18.7	1.9	1.2	1.4

Minerals (mg)

Ca	P	Fe	Na	K
34	211	0.4	91	405

Vitamins (mg)

Thiamine	Riboflavin	Niacin	Ascorbic Acid
0.08	0.14	2.7	0

Source: Data from Siong et al. (1987).

Scomber japonicus Houttuyn, 1782

Order: Perciformes

Family: Scombridae

Common name: Chub Mackerel

Distribution: Indo-Pacific

Habitat: Pelagic-neritic; oceanodromous; depth range 0–300 m

Description: This species grows to a maximum length of 64 cm and weight of 2.9 kg. Swim bladder is present. Anal fin spine is conspicuous, clearly separated from anal rays but joined to them by a membrane. Back has narrow stripes that zigzag and undulate. Belly is unmarked or has wavy lines. Caudal peduncle has five finlets on the upper and lower edges.

Nutritional Facts (per 56 g)

Calories	90
Calories from fat	35
Total fat	4.0 g
Saturated fat	1.0 g
Trans fat	0 g
Cholesterol	30 mg
Sodium	150 mg
Total carbohydrate	<1 g
Protein	14.0 g

Source: Data from DailyBurn Tracker (2014).

Proximate Composition (Values Based on Percentage of Wet Mass)

Moisture (%)	Ash (%)	Lipid (%)	Protein (%)	Energetic Value (kJ/g Wet Mass)
69.66	2.62	8.13	16.23	7.06

Source: Data from Eder and Lewis (2005).

Rastrelliger kanagurta (Cuvier, 1816)

Order: Perciformes

Family: Scombridae

Common name: Indian mackerel

Distribution: Indo-West Pacific; eastern Mediterranean

Habitat: Pelagic-neritic; oceanodromous; depth range 20–90 m

Description: This species grows to a maximum length of 38 cm. Head is longer than body depth. A black spot is on the body near the lower margin of the pectoral fin. Swim bladder is present. Anal spine is rudimentary.

Nutritional Facts

Proximate Composition (per 100 g of Edible Portion)

Energy (Kcal)	Moisture (%)	Protein (g)	Fat (g)	Carbohydrate (g)	Ash (g)
136	71.2	22.4	5.2	0	1.4

Minerals (mg)

Ca	P	Fe	Na	K	Mg	Co	Cu	Mn	Zn
0.59	88.68	11.87	464.47	27	268	1.3	59	370	710

Source: Data from Siong et al. (1987) and Nurnadfia (2013).

Vitamins (mg)

Thiamine	Riboflavin	Niacin	Ascorbic Acid
0.05	0.15	4.9	0

Source: Data from Siong et al. (1987).

Fatty Acids

EPA (µg/g)	DHA (µg/g)
17.7	67.5

EPA: eicosapentaenoic acid; DHA: docosahexaenoic acid.

Source: Data from Wan Rosli et al. (2012).

Fat (g/100 g)	4.54
Cholesterol (mg/100 g)	49.1
Saturated fatty acid	7.7 (%)
Monounsaturated	3.3 (%)
Polyunsaturated (ω6)	20.0 (%)
(ω3)	33.4 (%)
Other polyunsaturated	35.3 (%)

Source: Data from Osman et al. (2001).

Scomberomorus cavalla (Cuvier, 1829)

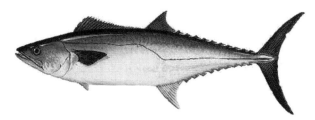

Order: Perciformes

Family: Scombridae

Common name: King mackerel

Distribution: Western Atlantic

Habitat: Reef-associated; oceanodromous; depth range 5–140 m

Description: This species grows to a maximum length of 184 cm and weight of 45 kg. Lateral line is abruptly curving downward below the second dorsal fin.

Intestine has two folds and three limbs. Adults have no black area on the anterior part of the first dorsal fin. Body is entirely covered with scales.

Nutritional Facts

Proximate Composition

Moisture (%)	Protein (%)	Oil (%)	Ash (%)
76	21	1.7	1.4

Source: Data from Stansby (1987).

Thunnus thynnus (Linnaeus, 1758)

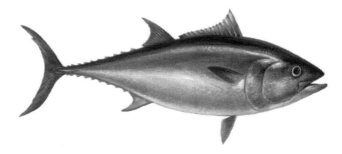

Order: Perciformes

Family: Scombridae

Common name: Bluefin tuna

Distribution: Western Atlantic and Eastern Atlantic

Habitat: Marine; brackish; pelagic-oceanic; oceanodromous; depth range 0–985 m

Description: This species grows to a maximum length of 4.5 m and weight of 684 kg. Body is deepest near the middle of the first dorsal fin base. The second dorsal fin is higher than the first. The pectoral fins are very short, less than 80% of the head length. Swim bladder is present. Lower sides and belly are silvery white, with colorless transverse lines alternated with rows of colourless dots. The first dorsal fin is yellow or bluish, and the second is reddish-brown. The anal fin and finlets are dusky yellow, and they are edged with black.

Nutritional Facts (per 100 g)

Water	68.090 g
Energy	144.000 Kcal
Energy	602.000 kJ
Protein	23.330 g
Total lipid (fat)	4.900 g
Ash	1.180 g
Calcium, Ca	8.000 mg
Iron, Fe	1.020 mg
Magnesium, Mg	50.000 mg
Phosphorus, P	254.000 mg
Potassium, K	252.000 mg
Sodium, Na	39.000 mg
Zinc, Zn	0.600 mg
Copper, Cu	0.086 mg

Continued

Nutritional Facts (per 100 g)

Manganese, Mn	0.015 mg
Selenium, Se	36.500 µg
Thiamin	0.241 mg
Riboflavin	0.251 mg
Niacin	8.654 mg
Pantothenic acid	1.054 mg
Vitamin B$_6$	0.455 mg
Folate, total	2.000 µg
Folate, food	2.000 µg
Folate, DFE	2.000 µg
Choline, total	65.000 mg
Vitamin B$_{12}$	9.430 µg
Vitamin A, IU	2183.000 IU

Vitamin A, RAE	655.000 µg
Retinol	655.000 µg
Vitamin E (alpha-tocopherol)	1.000 mg
Fatty acids, total saturated	1.257 g

DFE: dietary folate equivalent; RAE: retinol activity equivalent.

Source: Data from *Titi Tudorancea Bulletin* (2014).

Proximate Composition

Moisture (%)	Protein (%)	Oil (%)	Ash (%)
70.3	24.2	5.7	1.3

Source: Data from Stansby (1987).

Thunnus albacares (Bonnaterre, 1788)

Order: Perciformes

Family: Scombridae

Common name: Yellowfin tuna

Distribution: Worldwide in tropical and subtropical seas, but absent from the Mediterranean Sea

Habitat: Marine; brackish; pelagic-oceanic; oceanodromous; depth range 1–250 m

Description: This species grows to a maximum length of 2.4 m and weight of 200 kg. It has a very long second dorsal fin and anal fin. The pectoral fin is moderately long, usually reaching beyond the second dorsal fin origin but not beyond the end of its base. Color is black to metallic dark blue, changing through yellow to silver on the belly. The belly frequently has about 20 broken, nearly vertical lines. The dorsal and anal fins and finlets are bright yellow.

Nutritional Facts

Proximate Composition (Dry Weight Basis)

Protein (%)	Carbohydrate (%)	Lipid (%)	Ash (%)
13.7	5.6	3.0	1.6

Vitamins (mg/100 g)

B$_1$	B$_2$	B$_3$	B$_5$	B$_6$	FA	A	K	D$_3$
0.61	0.59	0.37	0.27	0.49	0.77	0.61	0.36	0.77

FA: folic acid.

Source: Data from Dhaneesh et al. (2012).

Proximate Composition (Wet Weight Basis)

Moisture (%)	Protein (%)	Oil (%)	Ash (%)
70.3	24.6	1.3	1.6

Source: Data from Stansby (1987).

Thunnus alalunga (Bonnaterre, 1788)

Order: Perciformes

Family: Scombridae

Common name: Albacore

Distribution: Cosmopolitan in tropical and temperate waters of all oceans

Habitat: Pelagic-oceanic; oceanodromous; depth range 0–600 m

Description: This species grows to a maximum length of 1.4 m. Anterior spines are much higher than posterior spines, giving the fin a strongly concave outline. Body has very small scales. Pectoral fins are remarkably long.

Nutritional Facts

Proximate Composition

Moisture (%)	Protein (%)	Oil (%)	Ash (%)
66.4	25.9	8.1	1.25

Source: Data from Stansby (1987).

Euthynnus affinis (Cantor, 1849)

Order: Perciformes

Family: Scombridae

Common name: Large tuna

Distribution: Indo-West Pacific

Habitat: Pelagic-neritic; oceanodromous; depth range 0–200 m

Description: This species grows to a maximum length of 1.0 m and weight of 14 kg. Anterior spines of first dorsal fin are much higher than those midway. Body is naked except for corselet and lateral line. Posterior portion of the back has a pattern of broken oblique stripes.

Nutritional Facts

Proximate Composition (per 100 g of Edible Portion)

Energy (Kcal)	Moisture (%)	Protein (g)	Fat (g)	Carbohydrate (g)	Ash (g)
117	72.7	23.2	2.7	0	1.4

Minerals (mg)

Ca	P	Fe	Na	K
20	273	1.6	51	344

Vitamins (mg)

Thiamine	Riboflavin	Niacin	Ascorbic Acid
0.10	0.22	7.1	0

Source: Data from Siong et al. (1987).

Fatty Acids

EPA (µg/g)	DHA (µg/g)
22.8	88.6

EPA: eicosapentaenoic acid; DHA: docosahexaenoic acid.

Source: Data from Wan Rosli et al. (2012).

Scomber australasicus Cuvier 1832

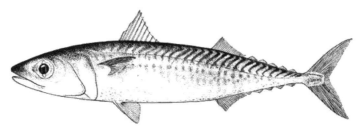

Order: Perciformes

Family: Scombridae

Common name: Blue mackerel

Distribution: Indo-West Pacific

Habitat: Pelagic-neritic; oceanodromous; depth range 87–200 m

Description: This species grows to a maximum length of 44 cm and weight of 1.4 kg. Body is covered with rather small scales. Palatine is narrow. Anal fin origin is clearly more posterior than that of second dorsal fin. Anal fin spine is independent from anal fin. Swim bladder is present. Snout is pointed.

Back has narrow oblique lines that zigzag and undulate. The belly is pearly white and marked with thin, wavy broken lines.

Nutritional Facts (per 2 oz.)

Calories	126	Sodium	235 mg
Total fat	8 g	Potassium	260 mg
Saturated	2 g	Protein	13 g
Polyunsaturated	2 g	Calcium	1%
Monounsaturated	3 g	Iron	4%
Cholesterol	43 mg		
Vitamin A	0%		
Vitamin C	3%		

Source: Data from All-Fish-Seafood-Recipes. com (2014).

Scomber scombrus Linnaeus, 1758

Order: Perciformes

Family: Scombridae

Common name: Atlantic mackerel

Distribution: North Atlantic, including the Mediterranean

Habitat: Marine; brackish; pelagic-neritic; oceanodromous; depth range 0–1000 m

Description: This species grows to a maximum length of 60 cm and weight of 3.4 kg. This species has no well-developed corselet. Interpelvic process is small and single. Anal fin spine is conspicuous, joined to the fin by a membrane but clearly independent of it.

Nutritional Facts (per 100 g)

Water	63.550 g
Energy	205.000 Kcal
Energy	858.000 kJ
Protein	18.600 g
Total lipid (fat)	13.890 g
Ash	1.350 g
Calcium, Ca	12.000 mg
Iron, Fe	1.630 mg
Magnesium, Mg	76.000 mg
Phosphorus, P	217.000 mg
Potassium, K	314.000 mg
Sodium, Na	90.000 mg
Zinc, Zn	0.630 mg
Copper, Cu	0.073 mg
Manganese, Mn	0.015 mg
Selenium, Se	44.100 µg
Vitamin C, total ascorbic acid	0.400 mg
Thiamin	0.176 mg
Riboflavin	0.312 mg
Niacin	9.080 mg
Pantothenic acid	0.856 mg
Vitamin B_6	0.399 mg
Folate, total	1.000 µg
Folate, food	1.000 µg
Folate, DFE	1.000 µg

Continued

Nutritional Facts (per 100 g)

Choline, total	65.000 mg
Vitamin B_{12}	8.710 µg
Vitamin A, IU	167.000 IU
Vitamin A, RAE	50.000 µg
Retinol	50.000 µg
Vitamin E (alpha-tocopherol)	1.520 mg
Vitamin D	360.000 IU
Vitamin K (phylloquinone)	5.000 µg
Fatty acids, total saturated	3.257 g
Fatty acids, total monounsaturated	5.456 g
Fatty acids, total polyunsaturated	3.350 g
Cholesterol	70.000 mg
Tryptophan	0.208 g
Threonine	0.815 g
Isoleucine	0.857 g
Leucine	1.512 g
Lysine	1.708 g
Methionine	0.551 g
Cystine	0.199 g
Phenylalanine	0.726 g
Tyrosine	0.628 g
Valine	0.958 g
Arginine	1.113 g
Histidine	0.548 g
Alanine	1.125 g
Aspartic acid	1.905 g
Glutamic acid	2.777 g
Glycine	0.893 g
Proline	0.658 g
Serine	0.759 g

DFE: dietary folate equivalent; RAE: retinol activity equivalent.

Source: Data from *Titi Tudorance Bulletin* (2013).

Scomberomorus guttatus (Bloch & Schneider, 1801)

Order: Perciformes

Family: Scombridae

Common name: Indo-Pacific king mackerel

Distribution: Indo-West Pacific

Habitat: Marine; brackish; pelagic-neritic; oceanodromous; depth range 15–200 m

Description: This species grows to a maximum length of 76 cm. Body is entirely covered with small scales. Lateral line has many auxiliary branches extending dorsally and ventrally in anterior third, curving down toward caudal peduncle. Sides are silvery white with several rows of round dark brownish spots scattered in three irregular rows along the lateral line. First dorsal fin membrane is black.

Nutritional Facts

Proximate Composition

Moisture (%)	Protein (%)	Oil (%)	Ash (%)
72.8	16.0	7.5	3.8

Source: Data from Moini et al. (2012).

Fatty Acids

EPA (µg/g)	DHA (µg/g)
33.7	118.0

EPA: eicosapentaenoic acid; DHA: docosahexaenoic acid.
Source: Data from Wan Rosli et al. (2012).

Scomberomorus commerson (Lacepède, 1800)

Order: Perciformes

Family: Scombridae

Common name: Narrow-barred Spanish mackerel

Distribution: Indo-West Pacific

Habitat: Pelagic-neritic; oceanodromous; depth range 10–70 m

Description: This species grows to a maximum length of 240 cm and weight of 70 kg. Lateral line is abruptly bent downward below the end of the second

199

dorsal fin. Vertical bars on trunk sometimes break up into spots ventrally. Middle third of first dorsal fin is white, and rest of fin is black.

Nutritional Facts

Proximate Composition (per 100 g of Edible Portion)

Energy (Kcal)	Moisture (%)	Protein (g)	Fat (g)	Carbohydrate (g)	Ash (g)
109	75.5	21.5	2.6	0	1.4

Minerals (mg)

Ca	P	Fe	Na	K	Mg	Co	Cu	Mn	Zn
0.34	60.50	9.31	227.27	25	243	0.5	67	324	874

Source: Data from Siong et al. (1987); and Nurnadia et al. (2013).

Vitamins (mg)

Thiamine	Riboflavin	Niacin	Ascorbic Acid
0.06	0.31	3.2	0

Source: Data from Siong et al. (1987).

Fatty Acids

Fat (g/100 g)	1.46
Cholesterol (mg/100 g)	41.1
Saturated fatty acid	6.4 (%)
Monounsaturated	4.1 (%)
Polyunsaturated ($\omega6$)	11.19 (%)
Polyunsaturated ($\omega3$)	47.9 (%)
Other polyunsaturated	32.8 (%)

Source: Data from Osman et al. (2001).

Acanthistius brasilianus (Cuvier, 1828)

Order: Perciformes

Family: Serranidae

Common name: Argentine seabass

Distribution: Southwest Atlantic

Habitat: Marine; brackish; benthopelagic

Description: This species grows to a maximum length of 60 cm. The total number of dorsal spines, dorsal soft rays, anal spines, and anal soft rays is 13, 15, 3, and 8, respectively. This species prefers the continental shelf in colder waters.

Nutritional Facts

Proximate Composition (Values Based on Percentage of Wet Mass)

Moisture (%)	Ash (%)	Lipid (%)	Protein (%)	Energetic Value (kJ/g Wet Mass)
76.36	2.79	2.27	12.41	3.82

Source: Data from Eder and Lewis (2005).

Epinephelus areolatus (Forsskål, 1775)

Order: Perciformes

Family: Serranidae

Common name: Areolate grouper

Distribution: Indo-Pacific

Habitat: Reef-associated; depth range 6–200 m

Description: This species grows to a maximum length of 47 cm and weight of 1.4 kg. It has gray to whitish color with numerous close-set orange to brown spots. A narrow white margin is on the tail. Body scales are ctenoid, and cycloid scales are seen on the thorax and ventrally on abdomen. Body has auxiliary scales. Body is moderately elongated and has auxiliary scales. Caudal fin is truncate or slightly emarginated.

Nutritional Facts

Proximate Composition

Moisture (%)	Protein (%)	Fat (%)	Ash (%)	Carbohydrate (%)	Energy Value (Kcal/100 g)
78.99	16.84	0.85	3.79	0.05	75.21

Minerals (mg/100 g)

Ca	Fe	P
1452.41	28.46	326.96

Source: Data from Palanikumar et al. (2014).

Epinephelus sexfasciatus (Valenciennes, 1828)

Order: Perciformes

Family: Serranidae

Common name: Coral cod, sixbar grouper

Distribution: Western Central Pacific

Habitat: Reef associated; depth range 10–80 m

Description: This species grows to a maximum length of 40 cm. Pectoral fins are not fleshy. Color of head and body is pale grayish-brown, with five dark brown bars on the body and one on nape. Scattered pale spots may be present on body, and some faint small brown spots are often on the edges of the dark bars. The soft dorsal, caudal, and pelvic fins are dusky gray, and the pectoral fins are grayish or orange-red. The jaws and ventral parts of the head are sometimes pale reddish-brown.

Nutritional Facts

Proximate Composition (per 100 g of Edible Portion)

Energy (Kcal)	Moisture (%)	Protein (g)	Fat (g)	Carbohydrate (g)	Ash (g)
80	79.5	18.4	0.5	0.4	1.2

Vitamins (mg)

Thiamine	Riboflavin	Niacin	Ascorbic Acid
0.02	0.04	1.4	1.1

Source: Data from Siong et al. (1987).

Minerals (mg)

Ca	P	Fe	Na	K	Mg	Co	Cu	Mn	Zn
0.54	82.44	12.39	235.69	48	184	0.4	109	260	792

Source: Data from Siong et al. (1987) and Nurnadia et al. (2013).

Fatty Acids

EPA (µg/g)	DHA (µg/g)
33.1	165.2

EPA: eicosapentaenoic acid; DHA: docosahexaenoic acid.

Source: Data from Wan Rosli et al. (2012).

Epinephelus tauvina (Forsskål, 1775)

Order: Perciformes

Family: Serranidae

Common name: Greasy grouper

Distribution: Indo-Pacific

Habitat: Reef-associated; oceanodromous; depth range 1–300 m

Description: This species grows to a maximum length of 90 cm. Color of head and body is pale greenish-gray or brown, with round dark spots that vary from dull orange-red to dark brown. A large black blotch (or group of black spots) is often visible on the body at the base of the last four dorsal-fin spines. Five subvertical dark bars may be present on body. Body is elongated.

Nutritional Facts

Proximate Composition (per 100 g of Edible Portion)

Energy (Kcal)	Moisture (%)	Protein (g)	Fat (g)	Carbohydrate (g)	Ash (g)
82	79.1	19.5	0.4	0	1.2

Vitamins (mg)

Thiamine	Riboflavin	Niacin	Ascorbic Acid
0.03	0.10	1.6	0

Minerals (mg)

Ca	P	Fe	Na	K
64	180	0.5	76	344

Source: Data from Siong et al. (1987).

Siganus rivulatus Forsskål Niebuhr, 1775

Order: Perciformes

Family: Siganidae

Common name: Rabbitfish, marbled spinefoot

Distribution: Western Indian Ocean

Habitat: Marine; brackish; reef associated

Description: This species grows to a maximum length of 27 cm. Upper body is gray, green, or brownish and silvery below. Iris is iridescent silver or golden.

Body color patterns extend to the fins. Spines are slender, pungent, and venomous. Cheeks are scaled. Midline of thorax, isthmus, and midline of belly are without scales.

Nutritional Facts

Proximate Composition (%)

Protein	Lipid	Moisture	Ash
19.30	2.37	78.00	1.56

Fatty Acids (%)

SFA	MUFA	PUFA
45.76	22.94	31.18

SFA: saturated fatty acids; MUFA: monounsaturated fatty acids; PUFA: polyunsaturated fatty acids.

Minerals (mg/kg, Wet Weight Base)

Cr	0.17
Cu	0.30
Mn	0.12
Zn	6.02
Fe	8.65
Ca	416.81
Mg	392.90
Na	729.62
P	2183.61
K	2683.24

Source: Data from Öksüz et al. (2010).

Siganus javus (Linnaeus, 1766)

Order: Perciformes

Family: Siganidae

Common name: Streaked rabbit fish

Distribution: Indo-Pacific

Habitat: Marine; brackish; reef associated; oceanodromous; depth range 0–15 m

Description: This species grows to a maximum length of 53 cm. Body is bronze above and white on belly and thorax. Iris is light brown. Pectoral fins are hyaline, and pelvic fins are white. Slender and pungent dorsal spines are present. Anal spines are stout.

Nutritional Facts

Proximate Composition (per 100 g of Edible Portion)

Energy (Kcal)	Moisture (%)	Protein (g)	Fat (g)	Carbohydrate (g)	Ash (g)
90	77.4	20.2	0.8	0.4	1.2

Vitamins (mg)

Thiamine	Riboflavin	Niacin	Ascorbic Acid
0.13	0.44	3.4	0

Minerals (mg)

Ca	P	Fe	Na	K
23	203	0.3	66	342

Source: Data from Siong et al. (1987).

Sillago sihama (Forsskål, 1775)

Order: Perciformes

Family: Sillaginidae

Common name: Silver whiting

Distribution: Indo-West Pacific

Habitat: Marine; brackish; reef associated; amphidromous; depth range 0–60 m

Description: This species grows to a maximum length of 31 cm. Swim bladder has two anterior and two posterior extensions. The anterior extensions extend forward and diverge to terminate on each side of the basioccipital above the auditory capsule. The species has a low lateral line with about 70 scales.

Nutritional Facts

Proximate Composition (per 100 g of Edible Portion)

Energy (Kcal)	Moisture (%)	Protein (g)	Fat (g)	Carbohydrate (g)	Ash (g)
7	78.0	20.5	0.6	0	1.3

Vitamins (mg)

Thiamine	Riboflavin	Niacin	Ascorbic Acid
0.07	0.05	2.0	0

Minerals (mg)

Ca	P	Fe	Na	K
61	223	0.3	78	350

Source: Data from Siong et al. (1987).

Sillago maculata Quoy & Gaimard, 1824

Order: Perciformes

Family: Sillaginidae

Common name: Trumpeter whiting

Distribution: Western Pacific: endemic to Australia

Habitat: Marine; brackish; demersal; nonmigratory; depth range 0–50 m

Description: This species grows to a maximum length of 30 cm. Anterolateral extensions of swim bladder are recurved posteriorly to reach level of vent. Base of pectoral fin has a black spot, and back and sides have dark blotches. The upper and lower blotches are frequently joined, at least posteriorly. The upper blotches are generally larger.

Nutritional Facts

Proximate Composition (per 100 g of Edible Portion)

Energy (Kcal)	Moisture (%)	Protein (g)	Fat (g)	Carbohydrate (g)	Ash (g)
81	78.7	19.6	0.2	0.4	1.4

Vitamins (mg)

Thiamine	Riboflavin	Niacin	Ascorbic Acid
0.04	0.04	2.3	1.7

Minerals (mg)

Ca	P	Fe	Na	K
50	222	0.4	97	335

Source: Data from Siong et al. (1987).

Stenotomus caprinus Jordan & Gilbert, 1882

Order: Perciformes

Family: Sparidae

Common name: Longspine porgy

Distribution: Western Atlantic

Habitat: Demersal; depth range 5–185 m

Description: This species grows to a maximum length of 30 cm. Body is oval, deep, and compressed. Color of the body is silvery without markings. Head profile is straight except for bump around eyes. Mouth is small, and front teeth in both jaws are in close-set rows. The first two dorsal spines are very small, and the third to fifth dorsal spines are long and filamentous. Pectoral fins are long, reaching past the anal fin origin.

Nutritional Facts

Proximate Composition

Moisture (%)	Protein (%)	Fat (%)	Ash (%)
80	18	1.5	1.0

Source: Data from Stansby (1987).

Diplodus sargus (Linnaeus, 1758)

Order: Perciformes

Family: Sparidae

Common name: White sea bream

Distribution: Eastern Atlantic

Habitat: Marine; brackish; demersal; oceanodromous; depth range 0–50 m

Description: This species grows to a maximum length of 45 cm and weight of 1.9 kg. Body has five black and four gray vertical bands. Snout is longer than the eye diameter.

Nutritional Facts

Proximate Composition

Moisture (%)	Protein (%)	Fat (%)	Ash (%)
76.24	20.38	1.41	1.65

Source: Data from Patrick et al. (2008).

Fatty Acids (% of Total Fatty Acids)

PUFA	DHA	EPA
47.2	28.4	3.9

PUFA: polyunsaturated fatty acids; DHA: docosahexaenoic acid; EPA: eicosapentaenoic acid.

Source: Data from Hossain (2011).

Sparus aurata Linnaeus, 1758

Order: Perciformes

Family: Sparidae

Common name: Gillhead sea bream

Distribution: Eastern Atlantic

Habitat: Marine; brackish; demersal; depth range 1–150 m

Description: This species grows to a maximum length of 70 cm and weight of 17.2 kg. Body is tall, with a large black spot on the gill cover. Snout is more than twice as long as the eye diameter. It is mainly carnivorous and accessorily herbivorous. It feeds on shellfish, including mussels and oysters.

Nutritional Facts

Proximate Composition

Moisture (%)	Protein (%)	Fat (%)	Ash (%)
76.5	21.2	0.9	1.5

Fatty Acids

Saturated	34.5%
Monounsaturated	27.5%
Polyunsaturated ω3	28.7%
EPA + DHA	24.6%

Source: Data from Nasopoulou et al. (2011).

Fatty Acids (% of Total Fatty Acids)

PUFA	DHA	EPA
19.6	10.3	4.6

PUFA: polyunsaturated fatty acids; DHA: docosahexaenoic acid; EPA: eicosapentaenoic acid.

Source: Data from Hossain (2011).

Sphyraena obtusata Cuvier, 1829

Order: Perciformes

Family: Sphyraenidae

Common name: Obtuse barracuda

Distribution: Indo-Pacific

Habitat: Marine; brackish; reef associated; depth range 20–120 m

Description: This species grows to a maximum length of 55 cm. Body is elongated and subcylindrical with small cycloid scales. Head is long and pointed. Mouth is large and horizontal; the tip of the lower jaw is protruding. First dorsal fin origin is slightly before the pectoral fin tip, the first spine equal to the second. Pelvic fins are well before the tip of the pectoral, closer to the anal than the tip of the lower jaw. Color is generally green above and silvery below.

Nutritional Facts

Proximate Composition (per 100 g of Edible Portion)

Energy (Kcal)	Moisture (%)	Protein (g)	Fat (g)	Carbohydrate (g)	Ash (g)
87	78.3	20.0	0.8	0	1.4

Vitamins (mg)

Thiamine	Riboflavin	Niacin	Ascorbic Acid
0.06	0.15	2.5	0

Minerals (mg)

Ca	P	Fe	Na	K
30	227	0.4	95	387

Source: Data from Siong et al. (1987).

Stromateus brasiliensis Fowler, 1906

Order: Perciformes

Family: Stromateidae

Common name: Southwest Atlantic butterfish

Distribution: Southwest Atlantic

Habitat: Benthopelagic; depth range 22–133 m

Description: This species grows to a maximum length of 38 cm. Body is blue-green dorsally and silvery white ventrally. Many round dark blue spots are present on the upper half of the body. Distal parts of fins are darker.

Nutritional Facts

Proximate Composition (Values Based on Percentage of Wet Mass)

Moisture (%)	Ash (%)	Lipid (%)	Protein (%)	Energetic Value (kJ/g Wet Mass)
67.25	1.86	17.48	12.48	9.92

Source: Data from Eder and Lewis (2005).

Pampus argenteus (Euphrasen, 1788)

Order: Perciformes

Family: Stromateidae

Common name: Silver pomfret

Distribution: Indo-West Pacific

Habitat: Benthopelagic; oceanodromous; depth range 5–110 m

Description: This species grows to a maximum length of 60 cm. Body is firm, very deep, oval, and compressed. Dorsal and anal fins are preceded by a series of 5–10 blade-like spines with anterior and posterior points. Pelvic fins are absent. Caudal fin is deeply forked; the lower lobe is longer than the upper. Color of the body is gray above, grading to silvery white toward the belly, with small black dots all over the body. Fins are faintly yellow, and vertical fins have dark edges.

Nutritional Facts

Proximate Composition (per 100 g of Edible Portion)

Energy (Kcal)	Moisture (%)	Protein (g)	Fat (g)	Carbo-hydrate (g)	Ash (g)
113	76.1	19.3	4.0	0	1.2

Vitamins (mg)

Thiamine	Riboflavin	Niacin	Ascorbic Acid
0.19	0.09	1.9	0.6

Source: Data from Siong et al. (1987).

Minerals (mg)

Ca	P	Fe	Na	K	Mg	Co	Cu	Mn	Zn
0.54	90.28	16.93	284.62	16	182	0.5	123	316	925

Source: Data from Siong et al. (1987) and Nurnadia et al. (2013).

Fatty Acids

EPA (µg/g)	DHA (µg/g)
29.9	83.7

EPA: eicosapentaenoic acid;
DHA: docosahexaenoic acid.
Source: Data from Wan Rosli et al. (2012).

Fat (g/100 g)	2.91
Cholesterol (mg/100 g)	40.3
Saturated fatty acid	9.2%
Monounsaturated	3.3%
Polyunsaturated (ω6)	13.5%
(ω3)	31.7%
Other polyunsaturated	36.3%

Source: Data from Osman et al. (2001).

Pampus chinensis (Euphrasen, 1788)

Order: Perciformes

Family: Stromateidae

Common name: Chinese silver pomfret

Distribution: Indo-West Pacific

Habitat: Marine; brackish; benthopelagic; amphidromous

Description: This species grows to a maximum length of 40 cm. Body is grayish to brownish dorsally, and silvery white on sides and covered in dark pigment spots. Fins are silvery to grayish, and darkest distally. Body is firm, very deep, and compressed. Caudal peduncle is short, deep, and strongly compressed. Eye is small, central, and much shorter than snout. Mouth is small, subterminal, and curved downward posteriorly. Scales are very small, cycloid, and deciduous, barely extending onto fin bases.

211

Nutritional Facts

Proximate Composition (per 100 g of Edible Portion)

Energy (Kcal)	Moisture (%)	Protein (g)	Fat (g)	Carbo-hydrate (g)	Ash (g)
90	78.4	17.7	2.8	0	1.3

Vitamins (mg)

Thiamine	Riboflavin	Niacin	Ascorbic Acid
0.60	0.10	1.9	0

Minerals (mg)

Ca	P	Fe	Na	K
28	47	0.2	153	283

Source: Data from Siong et al. (1987).

Trichiurus haumela (Forsskål, 1775) (= *Trichiurus lepturus*)

Order: Perciformes

Family: Trichiuridae

Common name: Largehead hairtail

Distribution: Circumtropical and temperate waters of the world

Habitat: Marine; brackish; benthopelagic; amphidromous; depth range 0–589 m

Description: This species grows to a common length of 234 cm and weight of 5 kg. Body is extremely elongated and compressed, and it tapers to a point. Mouth is large with a dermal process at the tip of each jaw. Dorsal fin is relatively high, and anal fin is reduced to minute spinules usually embedded in the skin or slightly breaking through. Pelvic and caudal fins are absent. Body is steely blue with silvery reflections.

Nutritional Facts

Proximate Composition (per 100 g of Edible Portion)

Energy (Kcal)	Moisture (%)	Protein (g)	Fat (g)	Carbohydrate (g)	Ash (g)
88	78.4	19.1	1.2	0.1	1.2

Vitamins (mg)

Thiamine	Riboflavin	Niacin	Ascorbic Acid
0.11	0.09	2.1	0

Minerals (mg)

Ca	P	Fe	Na	K
32	197	0.5	105	280

Source: Data from Siong et al. (1987).

Lepturacanthus savala (Cuvier, 1829)

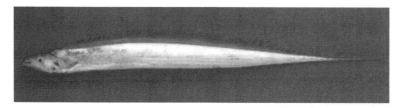

Order: Perciformes

Family: Trichiuridae

Common name: Savalai hairtail

Distribution: Indo-West Pacific

Habitat: Marine; brackish; benthopelagic; amphidromous

Description: This species grows to a common length of 100 cm. Pelvic and caudal fins are absent, and anal fin is reduced to spinules (about 75). Color of the body is steely blue with metallic reflections and the tapering part white. Tips of both jaws are black.

Nutritional Facts

Proximate Composition (Dryfish Values)

Moisture (%)	Protein (%)	Fat (%)	Carbohydrate (%)	Ash (%)
14.2	71.3	8.7	0	4.7

Source: Data from Siddique et al. (2012).

Xiphias gladius Linnaeus, 1758

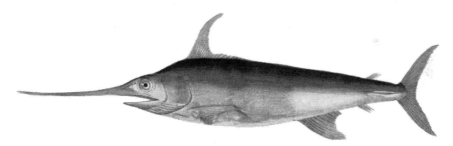

Order: Perciformes

Family: Xiphiidae

Common name: Swordfish

Distribution: Atlantic, Indian, and Pacific oceans

Habitat: Pelagic-oceanic; oceanodromous; depth range 0–800 m

Description: This species grows to a common length of 455 cm and weight of 650 kg. Body is blackish-brown fading to light brown below. The first dorsal fin has blackish-brown membrane, and other fins are brown or blackish-brown. A long, flat, sword-like bill is present, and pelvic fins are absent.

Nutritional Facts (per 100 g Cooked)

Total fat	8 g
Saturated fat	1.9 g
Polyunsaturated fat	1.4 g
Monounsaturated fat	3.5 g
Trans fat	0.1 g
Cholesterol	78 mg
Sodium	97 mg
Potassium	499 mg
Total carbohydrate	0 g
Dietary fiber	0 g
Sugar	0 g
Protein	23 g

Vitamin A	2%	Vitamin C	0%
Calcium	0%	Iron	2%
Vitamin D	166%	Vitamin B$_6$	30%
Vitamin B$_{12}$	26%	Magnesium	8%

Source: Data from USDA National Nutrient
Database for Standard Reference,
Release 26

Proximate Composition

Moisture (%)	Ash (%)	Lipid (%)	Protein (%)
76	1.3	3.7	19.9

Source: Data from Stansby (1987).

Iluocoetes fimbriatus Jenyns, 1842

Order: Perciformes

Family: Zoarcidae

Common name: Eelpout

Distribution: Southeast Pacific and Southwest Atlantic

Habitat: Marine; brackish; demersal; depth range 0–600 m

Description: This species grows to a maximum length of 36 cm. Dorsal-fin origin is associated with vertebrae 2–4. Well-developed oral valve is present. Gill rakers are blunt. Pelvic-fin rays are ensheathed. Head and pectoral-fin base and axil have no scales. Lateral line is mediolateral.

Nutritional Facts

Proximate Composition (Values Based on Percentage of Wet Mass)

Moisture (%)	Ash (%)	Lipid (%)	Protein (%)	Energetic Value (kJ/g Wet Mass)
78.15	3.00	3.22	8.82	3.36

Source: Data from Eder and Lewis (2005).

Mancopsetta maculata (Günther, 1880)

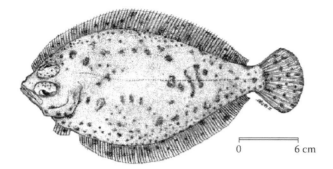

0   6 cm

Order: Pleuronectiformes

Family: Achiropsettidae

Common name: Antarctic armless flounder

Distribution: Southern Ocean, South Atlantic Ocean, and South Indian Ocean

Habitat: Demersal, depth range 100–1115 m; lower continental shelves

Description: This species grows to a maximum length of 50 cm. Dorsal soft rays and anal soft rays number 129 and 104, respectively. Pelvic fin on ocular side (with seven soft rays) is extremely larger than on blind side (with five soft rays), and caudal fin is rounded.

Nutritional Facts

Proximate Composition

Moisture (%)	Ash (%)	Lipid (%)	Protein (%)	Energetic Value (kJ/g Wet Mass)
60.55	1.06	11.29	27.90	11.45

Source: Data from Eder and Lewis (2005).

Cynoglossus senegalensis (Kaup, 1858)

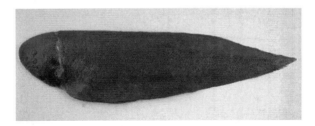

Order: Pleuronectiformes

Family: Cynoglossidae

Common name: Senegalese tonguesole

Distribution: Eastern Atlantic Ocean

Habitat: Marine; brackish; demersal; depth range 10–110 m

Description: This species grows to a maximum length of 66 cm. It is normally found on sand and mud bottoms of coastal waters and feeds on mollusks, shrimps, crabs, and fish.

Nutritional Facts

Proximate Composition

Moisture (%)	Protein (%)	Ash (%)	Fat (%)	Carbohydrate (%)	Energy (kJ/kg)
65.0	19.8	4.5	9.0	9.2	384.5

Minerals (mg/100 g)

Na	K	Ca	Mg	Fe	Zn	Cu	P	Mn	Pb
308.1	11.0	93.3	7.4	92.2	1.0	2.8	6.2	35.2	4.5

Source: Data from Udo and Arazu (2012).

Cynoglossus arel (Schneider Bloch 1801)

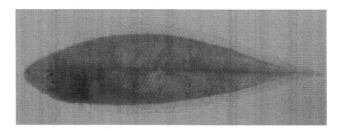

Order: Pleuronectiformes

Family: Cynoglossidae

Common name: Large-scale tongue sole

Distribution: Indo-West Pacific

Habitat: Marine; freshwater; brackish; demersal; amphidromous; depth range 9–125 m

Description: This species grows to a maximum length of 40 cm. Eyed side of this species is uniform brown, with a dark patch on gill cover; blind side is white. Body is elongate. Eyes have a small, scaly, interorbital space. Snout is obtusely pointed. Caudal-fin rays are usually 10. Midlateral-line scales are 56 to 70. Scales are large and ctenoid on eyed side of body. Scale rows between lateral lines on eyed side of body are 7–9.

Nutritional Facts

Proximate Composition

Moisture (%)	Protein (%)	Ash (%)	Fat (%)	Carbohydrate (%)	Energy (kJ/kg)
80.4	18.5	1.4	0.7	0	4.7

Source: Data from Nurnadia et al. (2011).

Minerals (mg/100 g)

Na	K	Ca	Mg	Co	Cu	Fe	Mn	Zn
0.3	43.2	175.6	38.2	182.8	36.7	10.2	65.2	682.6

Source: Data from Nurnadia et al. (2013).

Cynoglossus lingua Hamilton, 1822

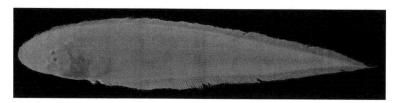

Order: Pleuronectiformes

Family: Cynoglossidae

Common name: Long tongue sole

Distribution: Indo-West Pacific

Habitat: Marine; freshwater; brackish; demersal; amphidromous; depth range 10–961 m

Description: This species grows to a maximum length of 45 cm. Eyed side is reddish-brown, sometimes with irregular brown-black patches, with a large black blotch on gill cover. Body is very elongate. Eyes have a small interorbital space. Number of caudal-fin rays is 10. No lateral lines are seen on blind side. Scales are comparatively large.

Nutritional Facts

Proximate Composition (per 100 g of Edible Portion)

Moisture (%)	Protein (g)	Fat (g)	Carbohydrate (g)	Ash (g)
80.0	18.0	0.4	0.3	1.3

Minerals (mg)

Ca	P	Fe	Na	K
33	151	0.4	230	206

Vitamins (mg)

Thiamine	Riboflavin	Niacin	Ascorbic Acid
0.01	0.14	1.0	0.8

Source: Data from Siong et al. (1987).

Paralichthys patagonicus Jordan, 1889

Order: Pleuronectiformes

Family: Paralichthyidae

Common name: Patagonian flounder

Distribution: Southeast Pacific

Habitat: Demersal; depth range 6–200 m

Description: This species grows to a maximum length of 48 cm. The maximum number of dorsal soft rays and anal soft rays is 84 and 65, respectively.

217

Scales are ctenoid only on the ocular side and cycloid on the blind side.

It feeds mainly on fishes, but also shrimps.

Nutritional Facts

Proximate Composition (Values Based on Percentage of Wet Mass)

Moisture (%)	Ash (%)	Lipid (%)	Protein (%)	Energetic Value (kJ/g Wet Mass)
76.17	4.05	2.27	12.11	3.77

Source: Data from Eder and Lewis (2005).

Hippoglossus hippoglossus (Linnaeus, 1758)

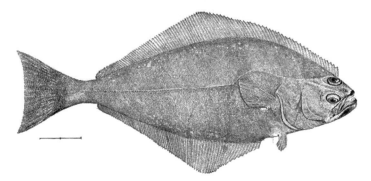

Order: Pleuronectiformes

Family: Pleuronectidae

Common name: Halibut

Distribution: Eastern Atlantic

Habitat: Demersal; oceanodromous; depth range 50–2000 m

Description: This species grows to a maximum length of 4.7 m and a weight of 320 kg. Body is uniformly dark brown or black. Young are marbled or spotted with paler marks. Adults are benthic but occasionally pelagic. They feed mainly on other fishes (cod, haddock, pogge, sand-eels, herring, and capelin), but also on cephalopods, large crustaceans, and other bottom-living animals.

Nutritional Facts (per 100 g)

Water	77.920 g
Energy	110.000 Kcal
Energy	460.000 kJ
Protein	20.810 g
Total lipid (fat)	2.290 g
Ash	1.360 g
Minerals	
Calcium, Ca	47.000 mg
Iron, Fe	0.840 mg
Magnesium, Mg	83.000 mg
Phosphorus, P	222.000 mg
Potassium, K	450.000 mg
Sodium, Na	54.000 mg
Zinc, Zn	0.420 mg
Copper, Cu	0.027 mg
Manganese, Mn	0.015 mg
Selenium, Se	36.500 µg

Vitamins	
Vitamin C, total ascorbic acid	
Thiamin	0.060 mg
Riboflavin	0.075 mg
Niacin	5.848 mg
Pantothenic acid	0.329 mg
Vitamin B$_6$	0.344 mg
Folate, total	12.000 µg
Folic acid	—
Folate, food	12.000 µg
Folate	12.000 µg
Choline, total	61.800 mg
Vitamin B$_{12}$	1.180 µg
Vitamin B$_{12}$, added	—
Vitamin A	157.000 IU
Vitamin A	47.000 µg
Retinol	47.000 µg
Vitamin E (alpha-tocopherol)	0.850 mg
Vitamin E, added	—

Vitamin K (phylloquinone)	0.100 µg
Fatty acids, total saturated	0.325 g
Fatty acids, total monounsaturated	0.750 g
Fatty acids, total polyunsaturated	0.730 g
Cholesterol	32.000 mg

DFE: dietary folate equivalent; RAE: retinol activity equivalent.

Source: Data from *Titi Tudorancea Bulletin* (2014).

Fatty Acids (% of Total Fatty Acids)

PUFA	DHA	EPA
43.9	25.4	12.2

PUFA: Polyunsaturated fatty acids; DHA: docosahexaenoic acid; EPA: eicosapentaenoic acid.
Source: Data from Hossain (2011).

Hippoglossus slenolepis Schmidt, 1904

Order: Pleuronectiformes

Family: Pleuronectidae

Common name: Pacific halibut

Distribution: North Pacific

Habitat: Demersal; oceanodromous; depth range 0–1200 m

Description: This species grows to a maximum length of 258 cm and a weight of 363 kg. Dorsal fin originates above anterior part of pupil in upper eye, generally low, higher in middle. Caudal fin is spread and slightly lunate. Pectorals are small.

Nutritional Facts

Proximate Composition

Moisture (%)	Ash (%)	Lipid (%)	Protein (%)
80	1.1	0.8	18.5

Source: Data from Stansby (1987).

Atheresthes stomias (Jordan & Gilbert, 1880)

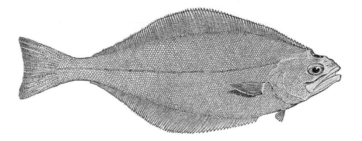

Order: Pleuronectiformes

Family: Pleuronectidae

Common name: Arrowtooth flounder

Distribution: North Pacific

Habitat: Demersal; depth range 18–950 m

Description: This species grows to a maximum length of 84 cm and a weight of 8.6 kg. The maximum number of dorsal soft rays and anal soft rays is 109 and 90, respectively. Dorsal fin originates over middle of upper eye. Caudal fin is slightly lunate. Pectorals are small.

Nutritional Facts (per 113 g)

Total fat	6 g
Saturated fat	0.5 g
Trans fat	0 g
Cholesterol	55 mg
Sodium	290 mg
Total carbohydrate	0 g
Dietary fiber	0 g
Sugars	0 g
Protein	17 g
Vitamin A	2%

Source: Data from Fatsecret (2014).

Proximate Composition

Moisture (%)	Ash (%)	Lipid (%)	Protein (%)
79.5	1.1	2.3	17.7

Source: Data from Krzynowek and Murphy (1987).

Pleuronectes platessa Linnaeus, 1758

220

Order: Pleuronectiformes

Family: Pleuronectidae

Common name: Plaice

Distribution: Northern Sea and Mediterranean Sea

Habitat: Marine; brackish; demersal; oceanodromous; depth range 0–200 m

Description: This species grows to a maximum length of 100 cm and a weight of 7 kg. Body is smooth with small scales. Bony ridge is seen behind the eyes. Upper side is brown or greenish-brown with irregularly distributed bright red or orange spots. The underside is white. Lateral line is straight, slightly curved above pectoral fin. Dorsal fin reaches the eye.

Vitamin B_{12}	1.59	µg
Vitamin C	1.5	mg
Sodium, Na	92	mg
Potassium, K	321	mg
Calcium, Ca	45	mg
Magnesium, Mg	22	mg
Phosphorus, P	220	mg
Iron, Fe	0.3	mg
Copper, Cu	0.052	mg
Zinc, Zn	0.68	mg
Iodine, I	34.0	µg
Manganese, Mn	0.04	mg
Chromium, Cr	1.61	µg
Selenium, Se	28.9	µg
Nickel, Ni	4.95	µg

Note: RE = retinol equivalents; NE = niacin equivalents (mg); α-TE = alpha-tocopherol equivalents (mg)

Nutritional Facts (per 100 g)

Energy	365	kJ
Protein, total	18.2	g
Total-N	2.9	g
Fat, total	1.5	g
Saturated fatty acids	0.2	g
Monounsaturated fatty acids	0.5	g
Polyunsaturated fatty acids	0.5	g
Ash	1.2	g
Moisture	79.3	g
Vitamin A	16.0	RE
Retinol	16	µg
ß-carotene eq.	0	µg
Vitamin D	3	µg
Vitamin E	0.8	α-TE
Alpha-tocopherol	0.8	mg
Vitamin K	0	µg
Vitamin B_1, thiamin	0.210	mg
Vitamin B_2, riboflavin	0.210	mg
Niacin equivalents	6.83	NE
Niacin	4.0	mg
Tryptophan	2.83	NE
Vitamin B_6	0.314	mg
Pantothenic Acid	0.80	mg
Biotin	1.2	µg
Folates	14	µg

Amino Acids	mg/100 g
Isoleucin	840
Leucine	1300
Lysine	1700
Methionine	550
Cystine	150
Phenylalanine	670
Tyrosine	610
Threonine	790
Tryptophan	170
Valine	990
Arginine	1200
Histidine	410
Alanine	1200
Aspartic acid	1800
Glutamic acid	2400
Glycine	1400
Proline	790
Serine	900
Cholesterol	60 mg

Source: Data from Department of Nutrition, Mørkhøj Bygade 19 – DK-2860. Jan. 13, (2009).

Psettodes erumei (Bloch & Schneider, 1801)

Order: Pleuronectiformes

Family: Psettodidae

Common name: Indian halibut

Distribution: Indo-West Pacific

Habitat: Demersal; depth range 1–100 m

Description: This species grows to a maximum length of 64 cm and a weight of 9 kg. Body is oval and flat, but thicker than in most other flatfishes. Mouth is large with strong teeth. Both eyes are on left or right side, and upper eye lies immediately below dorsal edge. Dorsal fin origin is well posterior to eyes. Anterior fin rays are spinous Lateral line is almost straight. Color of the body is usually brown or gray, sometimes with four broad, dark crossbars. Dorsal, anal, and caudal fin tips are black. Blind side is occasionally partially colored.

Nutritional Facts

Proximate Composition (per 100 g of Edible Portion)

Energy (Kcal)	Moisture (%)	Protein (g)	Fat (g)	Carbohydrate (g)	Ash (g)
81	78.6	19.4	0.2	0.5	1.3

Minerals (mg)

Ca	P	Fe	Na	K
43	216	0.4	101	420

Vitamins (mg)

Thiamine	Riboflavin	Niacin	Ascorbic Acid
0.04	0.06	3.4	0

Source: Data from Siong et al. (1987).

Scaophthalmus maximus (Linnaeus, 1758)

Order: Pleuronectiformes

Family: Scophthalmidae

Common name: Turbot

Distribution: Northeast Atlantic

Habitat: Marine; brackish; demersal; oceanodromous; depth range 20–70 m

Description: This species grows to a maximum length of 100 cm and weight of 25 kg. Body is almost circular. Eye side is without scales but with large bony tubercles. Adults live on sandy, rocky, or mixed bottoms. They are rather common in brackish waters and feed mainly on other bottom-living fishes (sand-eels, gobies, etc.), and also, to a lesser extent, on larger crustaceans and bivalves.

Nutritional Facts

Proximate Composition (per 100 g)

Total Fat	6 g
Saturated Fat	2 g
Trans Fat	0 g
Cholesterol	98 mg
Sodium	306 mg
Total carbohydrate	0 g
Protein	33 g
Vitamin A	1%
Vitamin C	0%
Calcium	4%
Iron	4%

Source: Data from SkipThePie.Org.

Fatty Acids (% of Total Fatty Acids)

PUFA	DHA	EPA
49.3	28.3	9.3

PUFA: polyunsaturated fatty acids; DHA: docosahexaenoic acid; EPA: eicosapentaenoic acid.
Source: Data from Hossain (2011).

Oncorhynchus kisutch (Walbaum, 1792)

Order: Salmoniformes

Family: Salmonidae

Common name: Coho salmon

Distribution: North Pacific Ocean

Habitat: Streams, small freshwater tributaries, estuarine, and marine waters

Description: This species grows to a maximum length of 61 cm and weight of 16 kg. Body is dark metallic blue or greenish-black, with silver sides and a light belly. Coho feeds on plankton and insects in freshwater, and switches to a diet of small fishes while in the ocean.

Nutritional Facts

Proximate Composition

Moisture (%)	Ash (%)	Lipid (%)	Protein (%)
69	1.2	8.3	21

Cholesterol (mg/100 g): 35.04.
Source: Data from Stansby (1987).

Oncorhynchus nerka (Walbaum, 1792)

Order: Salmoniformes

Family: Salmonidae

Common name: Sockeye salmon

Distribution: Pacific coast

Habitat: Lakes; estuarine and marine waters

Description: This species grows to a maximum length of 86 cm and weight of 3.6 kg. While spawning, they turn bright red with a green head; and while in the ocean, they are bluish-black with silver sides. In freshwater, they feed on aquatic insects and plankton; and in the ocean, they eat "amphipods," "copepods," squid, and some fishes.

224

Nutritional Facts (per 1 oz. of Boneless)

Total fat	2.43 g
Saturated fat	0.424 g
Polyunsaturated fat	0.533 g
Monounsaturated fat	1.169 g
Cholesterol	18 mg
Sodium	13 mg
Potassium	111 mg
Protein	6.04 g
Vitamin A	1%

Vitamin C	0%
Calcium	0%
Iron	1%

Source: Data from Fatsecret (2014).

Proximate Composition

Moisture (%)	Protein (%)	Oil (%)	Ash (%)
68	22	8.9	1.1

Source: Data from Stansby (1987).

Oncorhynchus mykiss (Walbaum, 1792) (= *Salmo gairdneri*)

Order: Salmoniformes

Family: Salmonidae

Common name: Steelhead (rainbow) trout

Distribution: Pacific Coast; Western Pacific; introduced worldwide

Habitat: Lakes; estuarine and marine waters

Description: This species grows to a maximum length of 120 cm and weight of 25 kg. Body is dark olive in color, shading to silvery-white on the underside, with a heavily speckled body with a pink-red stripe along the sides. In the ocean, they become more silver.

Nutritional Facts (per 100 g)

Water	72.730 g
Energy	138.000 Kcal
Energy	577.000 kJ
Protein	20.870 g
Total lipid (fat)	5.400 g
Ash	1.430 g

Calcium, Ca	67.000 mg
Iron, Fe	0.270 mg
Magnesium, Mg	32.000 mg
Phosphorus, P	282.000 mg
Potassium, K	451.000 mg
Sodium, Na	35.000 mg
Zinc, Zn	0.410 mg
Copper, Cu	0.046 mg
Manganese, Mn	0.018 mg
Selenium, Se	12.600 µg
Vitamin C, total ascorbic acid	2.900 mg
Thiamin	0.203 mg
Riboflavin	0.073 mg
Niacin	8.223 mg
Pantothenic acid	1.440 mg
Vitamin B_6	0.619 mg
Folate, total	11.000 µg
Choline, total	65.000 mg
Vitamin B_{12}	3.770 µg
Vitamin A	280.000 IU
Vitamin A	84.000 µg_RAE
Retinol	84.000 µg

Continued

225

Nutritional Facts (per 100 g)

Vitamin E (alpha-tocopherol)	0.030 mg
Vitamin K (phylloquinone)	0.100 μg
Fatty acids, total saturated	1.554 g
Cholesterol	59.000 mg

RAE: retinol activity equivalent.
Source: Data from *Titi Tudorancea Bulletin* (2014).

Proximate Composition

Moisture (%)	Protein (%)	Oil (%)	Ash (%)
68	21.2	9.5	1.3

Source: Data from Stansby (1987).

Micromesistius australis Norman, 1937

Order: Gadiformes

Family: Gadidae

Common name: New Zealand southern blue whiting

Distribution: Southwest Atlantic; southeast Pacific

Habitat: Marine; benthopelagic; oceanodromous; depth range 50–900 m

Description: This species grows to a maximum length of 90 cm and weight of 850 g. Dorsal soft rays and anal soft rays number 55 and 71, respectively. Maximum number of vertebrae in this species is 57. This species invades shelf waters during summer and concentrates over the continental slope in winter. It forms schools.

Nutritional Facts

Proximate Composition

Moisture (%)	Ash (%)	Lipid (%)	Protein (%)	Energetic Value (kJ/g Wet Mass)
73.88	2.47	5.55	13.42	5.38

Values based on percentage of wet mass.
Source: Data from Eder and Lewis (2005).

Minerals

Sodium (%)	Calcium (%)	Iron 1% (%)
26.2	12.8	1.0

Vitamins

A	C
7.8%	4.2%

Gadus morhua Linnaeus, 1758

Order: Gadiformes

Family: Gadidae

Common name: Atlantic cod

Distribution: North Atlantic; North Sea and the Barents Sea

Habitat: Coastal waters with depths of 500 to 600 m, and in the open ocean

Description: This species grows to a maximum length of 150 cm and weight of 40 kg. Head is rather disproportionately large for the body, with the upper jaw protruding over the lower jaw. The color of the body ranges from reddish or greenish where the water is populated by algae, and pale gray where the fish is found in deep water or near a sandy bottom. It has a barbel on the end of its chin. It has three dorsal and two anal fins, as in other members of the family.

Nutritional Facts (per 100 g)

Calories	82
Total fat	0.7 g
Saturated fat	0.1 g
Polyunsaturated fat	0.2 g
Monounsaturated fat	0.1 g
Cholesterol	43 mg
Sodium	54 mg
Potassium	413 mg
Total carbohydrate	0 g
Protein	18 g

Source: Data from USDA, National Nutrient Database for Standard Reference Release 26.

Gadus macrocephalus Tilesius, 1810

Order: Gadiformes

Family: Gadidae

Common name: Pacific cod

Distribution: North Pacific

227

Habitat: Marine; demersal; oceanodromous; depth range 0–1280 m

Description: This species grows to a maximum length of 120 cm and weight of 22 kg. It is distinguished by the presence of three dorsal and two anal fins, and a long chin barbel. Body is brown or gray dorsally, becoming paler ventrally. Dark spots or vermiculating patterns are present on the sides. Fins are somewhat dusky in color and are usually white-edged. Dorsal, anal, and caudal fins are with white edges.

Vitamins (%)

A	C	E	B$_{12}$	Folate
1	6	4	17	2

Minerals (%)

Na	Ca	K	Fe
3	1	13	2

Cholesterol: 14%.

Source: Data from SELFNutrition Data (2014).

Nutritional Facts

Proximate Composition

Moisture (%)	Protein (%)	Fat (%)	Ash (%)
81.4	18.0	0.6	1.2

Source: Data from Stansby (1987).

Melanogrammus aeglefinnus (Linnaeus, 1758)

Order: Gadiformes

Family: Gadidae

Common name: Haddock

Distribution: Both sides of the North Atlantic

Habitat: Depths of 40–300 m

Description: This species is easily recognized by a black lateral line running along its white side and a distinctive dark blotch above the pectoral fin, often described as a "thumbprint," the "devil's thumbprint," or "St. Peter's mark." It thrives in temperatures of 2 to 10°C (36 to 50°F). Juveniles prefer shallower waters, and larger adults prefer deeper waters.

Nutritional Facts

Proximate Composition

Moisture (%)	Protein (%)	Oil (%)	Ash (%)
81.7	17.2	0.25	1.23

Source: Data from Stansby (1987).

Pollachius virens (Linnaeus, 1758)

Order: Gadiformes

Family: Gadidae

Common name: Pollock

Distribution: Eastern Atlantic

Habitat: Demersal; oceanodromous; depth range 37–364 m

Description: This species grows to a maximum length of 130 cm and weight of 32 kg. Chin barbel is small. Lateral line is smooth along its entire length. Body color is brownish-green dorsally, becoming only slightly paler ventrally. The lateral line is pale.

Nutritional Facts (per 100 g)

Water	78.180 g
Energy	92.000 Kcal
Energy	385.000 kJ
Protein	19.440 g
Total lipid (fat)	0.980 g
Ash	1.410 g
Calcium, Ca	60.000 mg
Iron, Fe	0.460 mg
Magnesium, Mg	67.000 mg

Phosphorus, P	221.000 mg
Potassium, K	356.000 mg
Sodium, Na	86.000 mg
Zinc, Zn	0.470 mg
Copper, Cu	0.050 mg
Manganese, Mn	0.015 mg
Selenium, Se	36.500 µg
Thiamin	0.047 mg
Riboflavin	0.185 mg
Niacin	3.270 mg
Pantothenic acid	0.358 mg
Vitamin B_6	0.287 mg
Folate, total	3.000 µg
Choline, total	65.000 mg
Vitamin B_{12}	3.190 µg
Vitamin A, IU	37.000 IU
Vitamin E (alpha-tocopherol)	0.230 mg
Vitamin K (phylloquinone)	0.100 µg
Fatty acids, total saturated	0.135 g

Source: Data from *Titi Tudorancea Bulletin* (2014).

Proximate Composition

Moisture (%)	Protein (%)	Oil (%)	Ash (%)
80.3	18.6	0.6	1.16

Source: Data from Stansby (1987).

Theragra chalcogramma (Pallas, 1814)

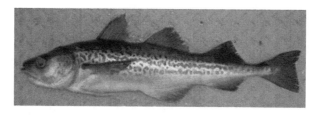

Order: Gadiformes

Family: Gadidae

Common name: Alaska Pollock

Distribution: North Pacific

Habitat: Marine; brackish; benthopelagic; nonmigratory; depth range 0–1280 m

Description: This species grows to a maximum length of 91 cm and weight of 4 kg. The dorsal fins are widely separated. The pelvic fins have a slightly elongated filament. The lateral line is continuous to about the back end of the first dorsal-fin base. On the head, there are lateral line pores. Body color is olive green to brown on the back and becomes silvery on the sides and pale ventrally, often with mottled patterns or blotches.

Nutritional Facts

Proximate Composition

Moisture (%)	Protein (%)	Oil (%)	Ash (%)
81.2	16.7	0.8	1.4

Cholesterol: 183 mg/100 g.
Source: Data from Stansby (1987).

Minerals (mg/100 g)

Na	Ca
123	9

Source: Data from Kevin Rail, Demand Media; and SFGate (2014).

Coelorhynchus fasciatus (Günther, 1878)

Order: Gadiformes

Family: Macrouridae

Common name: Banded rattail

Distribution: Atlantic, Indian, and Pacific Oceans

Habitat: Waters with a depth of 0–3650 m

Description: It has a large head, a short trunk, and a long tapering tail, which usually lacks a caudal fin. The first dorsal fin is usually high with the first rays spinous, the second dorsal fin is low, the pelvic fins are thoracic, and a chin barbel is usually present. Head is moderate or bulky, and mouth is terminal to inferior. Scales are cycloid, usually with spinules (but not on posterior edge). Maximum length of this species is 1.5 m.

Nutritional Facts

Proximate Composition (Values Based on % Wet Mass)

Moisture (%)	Ash (%)	Lipid (%)	Protein (%)	Energetic Value (kJ/g Wet Mass)
76.44	3.44	4.00	13.34	4.74

Source: Data from Eder and Lewis (2005).

Macruronus magellanicus Lönnberg, 1907

Order: Gadiformes

Family: Merlucciidae

Common name: Patagonian grenadier

Distribution: Southeast Pacific and Southwest Atlantic

Habitat: Benthopelagic; oceanodromous; depth range 30–500 m

Description: This species grows to a maximum length of 115 cm and weight of 5 kg. Dorsal part of body is purplish-blue. Belly is silvery with a slight bluish tinge. Small melanophores are found scattered on fin membrane of dorsal and anal fins. Inside of mouth is blackish.

Nutritional Facts

Proximate Composition (Values Based on Percentage of Wet Mass)

Moisture (%)	Ash (%)	Lipid (%)	Protein (%)	Energetic Value (kJ/g Wet Mass)
72.82	2.07	8.59	13.14	6.51

Source: Data from Eder and Lewis (2005).

Merluccius australis (Hutton, 1872)

Order: Gadiformes

Family: Merlucciidae

Common name: Southern hake

Distribution: Circumglobal in the southern hemisphere

Habitat: Benthopelagic; oceanodromous; depth range 28–1000 m

Description: This species grows to a maximum length of 155 cm. Body is more slender than other hakes. Pectoral fins are long and slender. A stripe reaches the anal fin in young individuals. Color of the body is steel gray on back, grading to silvery white ventrally.

231

Merluccius hubbsi Marini, 1933

Order: Gadiformes

Family: Merlucciidae

Common name: Argentine hake

Distribution: Southwest Atlantic

Habitat: Benthopelagic; oceanodromous; depth range 50–800 m

Description: This species grows to a maximum length of 95 cm. Pectoral fins are relatively short, not reaching level of anal fin origin. Color of the body is silvery with golden luster on back, silvery white on belly.

Nutritional Facts

Proximate Composition (Values Based on Percentage of Wet Mass)

	Moisture (%)	Ash (%)	Lipid (%)	Protein (%)	Energetic Value (kJ/g Wet Mass)
M. australis	76.08	2.03	4.48	15.66	5.48
M. hubbsi	77.27	2.84	3.65	14.65	4.91

Source: Data from Eder and Lewis (2005).

Salilota australis (Günther, 1878)

Order: Gadiformes

Family: Moridae

Common name: Tadpole codling

Distribution: Southeast Pacific and Southwest Atlantic

Habitat: Demersal; oceanodromous; depth range 30–1000 m

Description: This species grows to a maximum length of 27 cm. A small, variably shaped patch of teeth is present on the head of the vomer. Pectoral fin extends beyond the anal fin origin. Ventral light organ is present. Color of the body is uniformly brown. Fins are dark-edged. Caudal fin is rounded.

Notophycis marginata (Gunther, 1878) (= *Austrophycis marginata*)

Order: Gadiformes

Family: Moridae

Common name: Dwarf codling

Distribution: Worldwide

Habitat: Benthopelagic; depth range 1200 m

Description: It has two dorsal fins (rarely three), a single long-based anal fin (rarely two), and a separate caudal fin with a narrow caudal peduncle. Some species have a short barbel under the chin. All fins lack spines. Some species are bioluminescent and have a light organ on the belly just in front of the anus.

Nutritional Facts

Proximate Composition (Values Based on Percentage of Wet Mass)

	Moisture (%)	Ash (%)	Lipid (%)	Protein (%)	Energetic Value (kJ/g Wet Mass)
S. australis	77.40	3.15	2.62	14.39	4.44
A. marginata	73.48	3.95	4.15	16.26	4.91

Source: Data from Eder and Lewis (2005).

Anguilla rostrata Lesueur, 1821

Order: Anguilliformes

Family: Anguillidae

Common name: American eel

Distribution: Atlantic coast

Habitat: Freshwater, estuary, and ocean

Description: This species has a slender snakelike body that is covered with a mucous layer, which makes the eel appear to be naked and slimy despite the presence of minute scales. A long dorsal fin runs from the middle of the back and is continuous with a similar ventral fin. Pelvic fins are absent, and the pectoral fin (which is small) is found near the midline, following the head and gill covers. Coloration of the body varies from olive green, brown shading, to greenish-yellow and light gray or white on the belly.

Nutritional Facts

Proximate Composition

Moisture (%)	Protein (%)	Far (%)	Ash (%)
72	19	9	1

Source: Data from Krzynowek and Murphy (1987).

Bassanago albescens (Barnard, 1923)

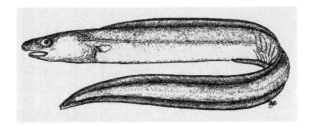

Order: Anguilliformes

Family: Congridae

Common name: Hairy conger

Distribution: Southeast Pacific

Habitat: Bathydemersal; depth range 270–1700 m

Description: It grows to a maximum length of 1 m. The numbers of dorsal spines, dorsal soft rays, anal spines, and anal soft rays are 0, 300, 0, and 213, respectively. Body is cream white, brownish dorsally and lighter below. Dorsal and anal fins have dark margin.

Nutritional Facts

Proximate Composition

Moisture (%)	Ash (%)	Lipid (%)	Protein (%)	Energetic Value (kJ/g Wet Mass)
75.78	2.58	4.76	14.14	5.25

Values based on percentage of wet mass.
Source: Data from Eder and Lewis (2005).

Tylosurus crocodilus crocodilus (Péron & Lesueur, 1821)

Order: Beloniformes

Family: Belonidae

Common name: Houndfish

Distribution: Indo-West Pacific

Habitat: Reef associated; oceanodromous; depth range 0–13 m

Description: This species, which grows to a maximum length of 150 cm, has a dark bluish back, silvery sides, and white ventrally. A distinct black lateral keel is present on caudal peduncle. Caudal fin is deeply forked. It has a relatively stout, cylindrical body and a shorter head as compared to other species of needle fishes.

Nutritional Facts

Proximate Composition (Dry Weight Basis)

Protein (%)	Carbohydrate (%)	Lipid (%)	Ash (%)
13.3	3.7	4.6	1.4

Vitamins (mg/100 g)

B$_1$	B$_2$	B$_3$	B$_5$	B$_6$	FA	A	K	D$_3$
0.39	0.48	0.41	0.31	0.34	0.79	0.72	0.19	0.44

FA: folic acid.

Source: Data from Dhaneesh et al. (2012).

Hyporhamphus dussumieri (Valenciennes, 1847)

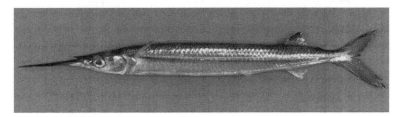

Order: Beloniformes

Family: Hemiramphidae

Common name: Dussumier's halfbeak

Distribution: Indo-Pacific

Habitat: Reef associated

Description: This species grows to a maximum length of 38 cm. It has a prolonged, beak-like lower jaw. Upper jaw is short, triangular, and scaly, and its width is 0.6–0.9 times in length. There are 14–16 dorsal and anal fin rays. Caudal fin is forked, with the lower lobe longer than the upper lobe.

Nutritional Facts

Proximate Composition (Dry Weight Basis)

Protein (%)	Carbohydrate (%)	Lipid (%)	Ash (%)
10.5	5.7	7.0	1.1

Vitamins (mg/100 g)

B$_1$	B$_2$	B$_3$	B$_5$	B$_6$	FA	A	K	D$_3$
0.39	0.86	0.51	0.37	0.43	0.84	0.68	0.67	0.54

FA: folic acid.

Source: Data from Dhaneesh et al. (2012).

Moolgarda seheli (Forsskål, 1775) (= *Valamugil seheli*)

Order: Mugiliformes

Family: Mugilidae

Common name: Bluespot mullet

Distribution: Indo-Pacific

Habitat: Marine; freshwater; brackish; reef-associated; catadromous

Description: This species grows to a maximum length of 60 cm and weight of 8 kg. Body is bluish-brown or green dorsally. Flanks and abdomen are silvery. Dusky spots are present on the upper row of scales, giving indistinct longitudinal stripes. Dorsal and upper lobe of caudal fin have dark-blue tips. Anal, pelvic, and pectoral fins are yellow. Pectorals also have dark blue spot dorsally at origin.

Nutritional Facts

Proximate Composition (per 100 g of Edible Portion)

Energy (Kcal)	Moisture (%)	Protein (g)	Fat (g)	Carbohydrate (g)	Ash (g)
135	72.8	19.4	6.6	0	1.2

Minerals (mg)

Ca	P	Fe	Na	K
35	206	0.9	58	323

Vitamins (mg)

Thiamine	Riboflavin	Niacin	Ascorbic Acid
0	0.43	3.0	0

Source: Data from Siong et al. (1987)

Mugil cephalus Linnaeus, 1758

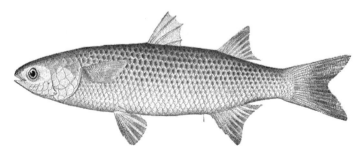

Order: Mugiliformes

Family: Mugilidae

Common name: Gray mullet

Distribution: Cosmopolitan in coastal waters of the tropical, subtropical, and temperate zones of all seas.

Habitat: Marine; freshwater; brackish; benthopelagic; catadromous; depth range 0–120 m

Description: This species grows to a maximum length of 100 cm. Color of the body is olive green dorsally. Sides are silvery shading to white ventrally. Lateral stripes are sometimes distinctive. Lips are thin. Pectoral fins are short (when folded forward, does not reach eye). Well-developed adipose eyelid is present.

Nutritional Facts

Proximate Composition

Moisture (%)	Protein (%)	Ash (%)	Fat (%)	Carbohydrate (%)	Energy (kJ/kg)
63.66	19.75	2.20	8.60	11.75	450.83

Minerals (mg/100 g)

Na	K	Ca	Mg	Fe	Zn	Cu	P	Mn	Pb
246.6	9.1	74.3	7.2	27.7	0.3	0.8	6.5	32.7	4.6

Source: Data from Udo and Arazu (2012).

Plotosus lineatus (Thunberg, 1787)

Order: Siluriformes

Family: Plotosidae

Common name: Striped sea catfish

Distribution: Indo-Pacific

Habitat: Marine; brackish; reef-associated; amphidromous; depth range 1–60 m

Description: This species grows to a maximum length of 32 cm. Dorsal and anal fins are continuous with caudal fin. Four pairs of mouth barbels are seen. A single highly venomous serrate spine is at the beginning of the first dorsal and each of the pectoral fins. These spines are dangerous and even fatal in rare cases.

Nutritional Facts

Fatty Acids

Fat (g/100 g)	2.79
Cholesterol (mg/100 g)	46.9
Saturated fatty acid	10.1 (%)
Monounsaturated	1.4
Polyunsaturated (ω6)	18.0
(ω3)	32.0
Other polyunsaturated	34.0
Total fatty acids	84.0

Source: Data from Osman et al. (2001).

Plotosus canius **Hamilton, 1822**

Order: Siluriformes

Family: Plotosidae

Common name: Gray eel-catfish

Distribution: Indo-West Pacific

Habitat: Marine; freshwater; brackish; demersal; amphidromous

Description: This species grows to a maximum length of 150 cm. It is a plain dusky-brown species with a black dorsal fin tip. It shows banded pattern at night. Its long barbels on the nostrils can reach pass the eyes. It feeds on crustaceans, mollusks, and fishes. Spines associated with anterior fins have potent venom.

Nutritional Facts

Proximate Composition (per 100 g of Edible Portion)

Energy (Kcal)	Moisture (%)	Protein (g)	Fat (g)	Carbohydrate (g)	Ash (g)
120	75.2	17.5	5.0	1.3	1.4

Vitamins (mg)

Thiamine	Riboflavin	Niacin	Ascorbic Acid
0.01	0.04	1.8	1.6

Source: Data from Siong et al. (1987).

Minerals (mg)

Ca	P	Fe	Na	K	Mg	Co	Cu	Mn	Zn
0.66	183.12	25.96	34	4.0314	77	0.8	76	275	716

Source: Data from Siong et al. (1987); and Nurnadia et al. (2013).

Netuma thalassina (Rüppell, 1837)

Order: Siluriformes

Family: Ariidae

Common name: Giant sea catfish

Distribution: Western Indian Ocean and Australia, Polynesia, and Japan

Habitat: Marine; freshwater; brackish; demersal; amphidromous; depth range 10–195 m

Description: This species grows to a common length of 70 cm and weight of 1 kg. Ariid catfish have a deeply forked caudal fin. Usually, three pairs of barbels are present. They possess some bony plates on their heads and near their dorsal fins. They feed mainly on crabs, prawns, and mantis shrimps (*Squilla* species) but also on fishes and mollusks.

Nutritional Facts

Proximate Composition (per 100 g of Edible Portion)

Energy (Kcal)	Moisture (%)	Protein (g)	Fat (g)	Carbohydrate (g)	Ash (g)
78	79.5	18.3	0.2	0	1.3

Vitamins (mg)

Thiamine	Riboflavin	Niacin	Ascorbic Acid
0.05	0.08	2.0	0

Minerals (mg)

Ca	P	Fe	Na	K
16	189	0.5	83	434

Source: Data from Siong et al. (1987).

Abalistes stellaris (Bloch & Schneider, 1801)

239

Order: Tetraodontiformes

Family: Balistidae

Common name: Starry triggerfish

Distribution: Indo-West Pacific

Habitat: Demersal; amphidromous

Description: This species grows to a maximum length of 60 cm. Scales are enlarged above the pectoral fin base, and just behind the gill slit a flexible tympanum is present. Scales of posterior body have prominent keels forming longitudinal ridges. A prominent groove in the skin extends anteriorly from front of eye for a distance of about one eye diameter. Caudal peduncle is depressed. Caudal fin rays of adults are prolonged above and below.

Nutritional Facts

Proximate Composition (per 100 g of Edible Portion)

Energy (Kcal)	Moisture (%)	Protein (g)	Fat (g)	Carbohydrate (g)	Ash (g)
75	80.1	18.5	0.4	0	1.4

Minerals (mg)

Ca	P	Fe	Na	K
23	80	0.7	173	311

Vitamins (mg)

Thiamine	Riboflavin	Niacin	Ascorbic Acid
0.1	0.03	2.9	0.6

Source: Data from Siong et al. (1987).

Congiopodus peruvianus (Cuvier, 1829)

Order: Scorpaeniformes

Family: Congiopodidae

Common name: Horsefish

Distribution: Southeast Pacific and Southwest Atlantic.

Habitat: Demersal

Description: It grows to a maximum length of 27 cm. Number of dorsal spines, dorsal soft rays, anal spines, and anal soft rays is 18, 15, 1, and 10, respectively.

Nutritional Facts

Proximate Composition

Moisture (%)	Ash (%)	Lipid (%)	Protein (%)	Energetic Value (kJ/g Wet Mass)
74.33	4.00	5.26	12.21	5.00

Source: Data from Eder and Lewis (2005).

240

Pleurogammus azonus Jordan & Metz, 1913

Order: Scorpaeniformes

Family: Hexagrammidae

Common name: Okhotsk atka mackerel

Distribution: Northwest Pacific

Habitat: Demersal; oceanodromous; depth range 0–240 m

Description: This species grows to a maximum length of 62 cm and weight of 1.6 kg. Dorsal fin is without notch; if present, it is very shallow. There are five lateral lines on the body. Strongly developed ridges are seen on the upper surface of the skull. Juveniles form large schools near the surface.

Nutritional Facts

Proximate Composition (%) (Based on 100 g Edible Weight)

Moisture (%)	Protein (%)	Fat (%)	Ash (%)
78	16.3	3.9	1.8

Source: Data from Bykov (1983).

Cottunculus granulosus Karrer, 1968

Order: Scorpaeniformes

Family: Psychrolutidae

Common name: Fathead

Distribution: Southwest Atlantic

Habitat: Bathydemersal; depth range 150–1250 m

Description: This species grows to a maximum length of 22 cm. Vomerine teeth are well developed. Bony plates on head and body are widely scattered, each with several prickles. Head spines are prominent and suborbital spines are distinct and are in three paired sets. Lateral line pores on body do not emerge in distinct tubes. Color pattern on body varies, with or without three irregular brownish bands.

Nutritional Facts

Proximate Composition (Values Based on Percentage of Wet Mass)

Moisture (%)	Ash (%)	Lipid (%)	Protein (%)	Energetic Value (kJ/g Wet Mass)
77.76	6.08	0.97	11.01	2.99

Source: Data from Eder and Lewis (2005).

Sebastes oculatus Valenciennes, 1833

Order: Scorpaeniformes

Family: Sebastidae

Common name: Patagonian redfish

Distribution: Southeast Pacific and Southwest Atlantic Oceans

Habitat: Demersal

Description: This species grows to a maximum length of 31 cm. The total number of dorsal spines, dorsal soft rays, anal spines, and anal soft rays is 13, 14, 3, and 7, respectively. Five dark blotches are seen dorsally. Four to five spots are present on lateral part of body. It is a viviparous species.

Nutritional Facts

Proximate Composition (Values Based on Percentage of Wet Mass)

Moisture (%)	Ash (%)	Lipid (%)	Protein (%)	Energetic Value (kJ/g Wet Mass)
73.88	3.85	2.99	16.61	5.12

Source: Data from Eder and Lewis (2005).

Prionotus nudigula Ginsburg, 1950

Order: Scorpaeniformes

Family: Triglidae

Common name: Red searobin

Distribution: Southwest Atlantic

Habitat: Demersal; depth range 15–200 m

Description: This species grows to a maximum length of 35 cm. Head is bony and casquelike. Pectoral fin has lower two or three rays, which are enlarged for food detection. Dorsal fins are separate.

Nutritional Facts

Proximate Composition (Values Based on Percentage of Wet Mass)

Moisture (%)	Ash (%)	Lipid (%)	Protein (%)	Energetic Value (kJ/g Wet Mass)
70.29	6.94	3.62	13.36	4.63

Source: Data from Eder and Lewis (2005).

Esox lucius Linnaeus, 1758

Order: Esociformes

Family: Esocidae

Common name: Northern pike

Distribution: North America; Eurasia

Habitat: Freshwater; brackish; demersal; potamodromous; depth range 0–30 m

Description: This species grows to a maximum length of 150 cm and weight of 28 kg. It has a long, flat, "duckbill" snout and large mouth with sharp teeth. Dorsal fin originates slightly in front of anal origin. Lateral line has 105–148 scales. Lateral line is notched posteriorly. Dorsal fin is located far to the rear, and anal fin is located under and arises a little behind dorsal. Pectorals and pelvic fins are low on body. Paired fins are rounded and paddle shaped.

Nutritional Facts (per 100 g)

Water	69.800 g
Energy	156.000 Kcal
Energy	651.000 kJ

Protein	16.600 g
Total lipid (fat)	8.000 g
Ash	1.300 g
Carbohydrate, by difference	4.300 g
Calcium, Ca	28.000 mg
Iron, Fe	2.100 mg
Phosphorus, P	412.000 mg
Thiamin	0.100 mg
Riboflavin	0.700 mg
Niacin	5.000 mg
Vitamin A, IU	860.000 IU

Source: Data from *Titi Tudorancea Bulletin* (2013).

Fatty Acids

Total Lipid	PUFA	DHA	EPA
0.2	58.3	28.0	6.7

Expressed as percentage of total fatty acids. PUFA: polyunsaturated fatty acids; DHA: docosahexaenoic acid; EPA: eicosapentaenoic acid.
Source: Data from Hossain (2011).

Genypterus blacodes (Forster, 1801)

Order: Ophidiiformes

Family: Ophidiidae

Common name: Pink ling, pink cusk-eel

Distribution: Southwest Pacific

Habitat: Bathydemersal; oceanodromous; depth range 22–1000 m

Description: This species grows to a maximum length of 200 cm and a weight of 25 kg. Body is pinkish-yellow, marbled, with irregular reddish-brown blotches dorsally. It feeds mainly on crustaceans such as Munida and scampi and also on fish. It is available all year around. It is oviparous, with oval pelagic eggs floating in a gelatinous mass.

Vitamin B$_2$	0.06 mg
Niacin	1.2 mg
Vitamin B$_6$	0.07 mg
Vitamin B$_{12}$	1.04 µg
Folate	3.2 µg
Pantothenic acid	0.34 mg
Vitamin C	0.8 mg
Sodium	112 mg
Potassium	272 mg
Calcium	37.6 mg
Magnesium	22.4 mg
Phosphorus	136 mg
Iron	0.24 mg
Zinc	0.4 mg
Copper	0.02 mg
Manganese	0.01 mg

Source: Data from CalorieSlism.

Nutritional Facts (per 80 g)

Energy	62 cal
Protein	14.56 g
Fat	0.08 g
Carbohydrate	0 g
Vitamin A retinol equivalent	4 µg
Vitamin E alpha tocopherol	0.16 mg
Vitamin B$_1$	0.02 mg

Proximate Composition (Values Based on Percentage of Wet Mass)

Moisture (%)	Ash (%)	Lipid (%)	Protein (%)	Energetic Value (kJ/g Wet Mass)
76.67	2.59	5.02	15.28	5.60

Source: Data from Eder and Lewis (2005).

Mallotus villosus (Müller, 1776)

Order: Osmeriformes

Family: Osmeridae

Common name: Capelin

Distribution: Circumpolar in the Arctic; North Atlantic; North Pacific

Habitat: Marine; freshwater; brackish; pelagic-oceanic; anadromous; depth range 0–725 m

Description: This species grows to a maximum length of 20 cm and a weight of 52 g. Adipose is with long base, about 1.5 times as long as the orbit or longer; outer margin is only slightly curved.

Body is olive green on dorsal surface, merging into silvery on sides and ventral surface.

Nutritional Facts (per 100 g)

Energy	770	kJ
Protein, total	9.1	g
Total-N	1.5	g
Fat, total	14.0	g
Carbohydrate total	5.7	g
Ash	6.2	g
Moisture	65.0	g
Cholesterol	312	

Source: Data from Department of Nutrition, Mørkhøj Bygade 19 – DK-2860 (2009).

Thaleichthys pacijicus (Richardson, 1836)

Order: Osmeriformes

Family: Osmeridae

Common name: Eulachon, Candlefish

Distribution: North Pacific

Habitat: Marine; freshwater; brackish; pelagic-neritic; anadromous; depth range 0–300 m

Description: This species grows to a maximum length of 34 cm. It is distinguished by the large canine teeth on the vomer and 18–23 rays in the anal fin. Adipose fin is sickle shaped. Paired fins are longer in males than in females. All fins have well-developed breeding tubercles in ripe males, which are poorly developed or absent in females. Adult coloration is brown to blue on back and top of the head. Sides are lighter to silvery white, and the ventral surface is white. Speckling is fine, sparse, and restricted to the back. Fins are transparent. Pectorals and caudal are often dusky.

Nutritional Facts

Proximate Composition

Moisture (%)	Ash (%)	Lipid (%)	Protein (%)
79.6	1.3	6.3	14.6

Source: Data from Krzynowek and Murphy (1987).

Saurida tumbil (Bloch, 1795)

Order: Aulopiformes

Family: Synodontidae

Common name: Greater lizardfish

Distribution: Indo-West Pacific

Habitat: Reef associated; amphidromous; depth range 10–60 m

Description: This species grows to a maximum length of 60 cm. Body is cigar shaped, rounded, or slightly compressed. The head is pointed and depressed, and the snout is broader than long. Color of the body is generally brown above and silver below. The back has faint cross-bands. The tips of the dorsal and pectorals, and the lower caudal lobe, are blackish.

Nutritional Facts

Proximate Composition

Moisture (%)	Protein (%)	Fat (%)	Ash (%)	Carbohydrate (%)	Energy (Kcal/100 g)
79.19	16.28	0.29	1.85	0.09	67.93

Minerals (mg/100 g)

Ca	Fe	P
475.26	0.05	321.48

Source: Data from Palanikumar et al. (2014).

Harpodon nehereus (Hamilton, 1822)

Order: Aulopiformes

Family: Synodontidae

Common name: Bombay duck

Distribution: Indo-West Pacific

Habitat: Marine; brackish; benthopelagic; oceanodromous

Description: This species grows to a maximum length of 40 cm. Scales are restricted to posterior half of the body. Posterior tip of pectoral fin reaches origin of pelvic fin. It spawns six batches of broods per year and is an aggressive predator.

Nutritional Facts

Proximate Composition (Dryfish Values)

Moisture (%)	Protein (%)	Fat (%)	Carbohydrate (%)	Ash (%)
24.9	57.1	7.5	2.2	6.7

Source: Data from Siddique et al. (2012).

Callorhynchus callorhynchus (Linnaeus, 1758)

Order: Chimaeriformes

Family: Callorhinchidae

Common name: Elephantfish, cockfish

Distribution: Southeast Pacific and Southwest Atlantic.

Habitat: Demersal; depth range 10–116 m

Description: This species grows to an average length of 89 cm. It has only one dorsal spine. Pectorals are very large. This species is abundant in sandy and muddy substrates, and it is apparently associated with *Stromateus brasiliensis* and *Discopyge tschudii*. It is an oviparous species.

Nutritional Facts

Proximate Composition (Values Based on Percentage of Wet Mass)

Moisture (%)	Ash (%)	Lipid (%)	Protein (%)	Energetic Value (kJ/g Wet Mass)
77.18	2.09	3.89	10.79	4.09

Source: Data from Eder and Lewis (2005).

ELASMOBRANCHS

Galeocerdo cuvieri (Péron & Lesueur, 1822)

Phylum: Chordata

Class: Chondrichthyes

Order: Carcharhiniformes

Family: Carcharhinidae

Common name: Tiger shark

Distribution: Circumglobal in tropical and temperate seas

Habitat: Marine; brackish; benthopelagic; oceanodromous; depth range 0–371 m

Description: This species grows to a common length of 7.5 m and weight of 807 kg. It is a huge, vertical tiger-striped shark with a broad, bluntly rounded snout; long upper labial furrows; and a big mouth with large, saw-edged, cockscomb-shaped teeth. Spiracles are present. Caudal keels are low. Body is gray above with vertical dark gray to black bars and spots that appear faded in adults, and white below.

Nutritional Facts

Proximate Composition

Moisture (%)	Ash (%)	Lipid (%)	Protein (%)
80	1.2	0.6	19.2

Source: Data from Stansby (1987).

Scoliodon sorrokowah Thillayampallam, 1928 (= *Scoliodon laticaudus*)

Order: Carcharhiniformes

Family: Scyliorhinidae

Common name: Indian dogfish

Distribution: Western Indo-Pacific

248

Habitat: Coastal waters (10–13 m deep), often close to rocky bottoms

Description: A small, stocky species, the spadenose shark has a broad head with a distinctive highly flattened, trowel-shaped snout. The eyes and nares are small. The corners of the mouth are well behind the eyes and have poorly developed furrows at the corners. Each tooth has a single, slender, blade-like, oblique cusp without serrations. The first dorsal fin is positioned closer to the pelvic than the pectoral fins, which are very short and broad. The second dorsal fin is much smaller than the anal fin. The back is bronze-gray in color, and the belly is white. The fins are plain but may be darker than the body. The maximum known length is 74 cm.

Nutritional Facts

Proximate Composition (per 100 g of Edible Portion)

Energy (Kcal)	Moisture (%)	Protein (g)	Fat (g)	Carbohydrate (g)	Ash (g)
104	75.6	25.6	0.2	0	1.4

Vitamins (mg)

Thiamine	Riboflavin	Niacin	Ascorbic Acid
0.04	0.07	3.3	1.2

Minerals (mg)

Ca	P	Fe	Na	K
13	219	0.4	74	300

Source: Data from Siong et al. (1987).

Schroederichthys bivius (Müller & Henle, 1838)

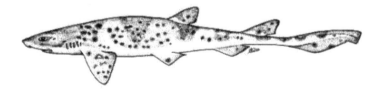

Order: Carcharhiniformes

Family: Scyliorhinidae

Common name: Narrowmouthed catshark

Distribution: Southeast Pacific and Southwest Atlantic Oceans

Habitat: Demersal; depth range 14–78 m

Description: This species grows to 70 cm. Trunk and tail are fairly slender in adults, but they are extremely attenuated in young. Snout is narrowly rounded. Anterior nasal flaps are narrow and lobate. Mouth is relatively narrow and long, especially in adult males. Color pattern is seven or eight dark brown saddles on gray-brown in dorsal surface.

Nutritional Facts

Proximate Composition (Values Based on Percentage of Wet Mass)

Moisture (%)	Ash (%)	Lipid (%)	Protein (%)	Energetic Value (kJ/g Wet Mass)
72.21	3.35	9.36	11.12	6.34

Source: Data from Eder and Lewis (2005).

Dasyatis zugei (Müller & Henle, 1841)

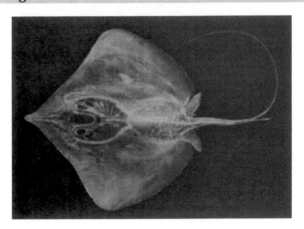

Order: Myliobatiformes

Family: Dasyatidae

Common name: Pale-edged stingray

Distribution: Indo-West Pacific

Habitat: Marine; brackish; demersal; amphidromous

Description: This species has a maximum length of 29 cm and common length of 18 cm. It feeds on bottom-dwelling organisms, primarily small crustaceans, as well as small fishes. It is an ovoviviparous species.

Nutritional Facts

Proximate Composition (per 100 g of Edible Portion)

Energy (Kcal)	Moisture (%)	Protein (g)	Fat (g)	Carbohydrate (g)	Ash (g)
100	76.1	24.2	0.3	0	1.3

Vitamins (mg)

Thiamine	Riboflavin	Niacin	Ascorbic Acid
0.04	0.06	2.9	0

Minerals (mg)

Ca	P	Fe	Na	K
16	106	0.7	87	240

Source: Data from Siong et al. (1987).

Gymnura poecilura (Shaw, 1804)

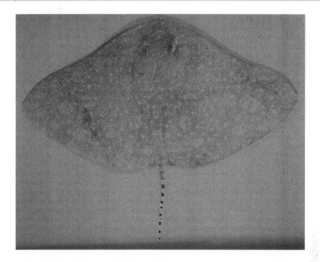

Order: Myliobatiformes

Family: Gymnuridae

Common name: Longtail butterfly ray

Distribution: Indo-Pacific

Habitat: Demersal

Description: This species grows to a common length of 2.5 m and width of 0.9 m. The pectoral fin disc of this species has the lozenge-shape characteristic of its family, measuring around twice as wide as long. The leading margin of the disc is gently sinuous, the trailing margin is convex, and the outer corners are mildly angular. The snout is short and broad, with a tiny protruding tip. The medium-sized eyes have larger, smooth-rimmed spiracles behind. The large mouth forms a transverse curve. The pelvic fins are small and rounded. The thread-like tail lacks dorsal or caudal fins. This species is brown to greenish-brown to gray above, with many small pale spots and sometimes also a smattering of dark dots. The underside is white, darkening at the edges of the fins. The tail has 9–12 black bands alternating with white bands.

Nutritional Facts

Proximate Composition

Energetic Value (kJ/g)	Moisture (%)	Protein (%)	Fat (%)	Carbohydrate (%)	Ash (%)
5.6	78.0	22.2	0.9	0	2.1

Source: Data from Nurnadia et al. (2011).

Minerals (mg/100 g)

Na	K	Ca	Mg	Co	Cu	Fe	Mn	Zn
0.71	181.91	336.96	28.78	233.19	76.76	12.64	127.59	964.04

Source: Data from Nurnadia et al. (2013).

Dipturus chilensis Guichenot, 1848

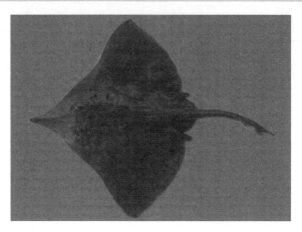

Order: Rajiformes

Family: Rajidae

Common name: Yellownose skate

Distribution: Southwest Atlantic: Uruguay, Argentina and around the Falkland/Malvinas Islands.

Habitat: soft bottoms

Description: Size of this species is 170 cm. Disc of this species is rhomboidal, and its snout is long, prominent, and stiff, with concave sides. Between 25 and 35 rows of teeth are present on the top jaw. Dorsal fins are close together, and there is no tail fin. Three to five spines are seen on the front and top of the eye. One large solitary spine is present on the midnape. Middle of the back may have small denticulations. Tail has 10–21 thorns from the region above the origin of pelvic fins rearwards. One or two small spines are present at the edge of the disc between the snout and pectoral. Body is gray above, with a few whitish and red-brown blotches, and a pair of larger, darker red-brown blotches on each pectoral fin. Belly is white with gray blotches or gray with white blotches, with dark pores.

Psammobatis scobina (Philippi, 1857)

Order: Rajiformes

Family: Arhynchobatidae

Common name: Raspthorn sandskate

Distribution: Southeast Pacific

Habitat: Demersal; depth range 40–450 m

Description: This is an oviparous species. Young may tend to follow large objects, such as their mother. Eggs are oblong capsules with stiff pointed horns at the corners deposited in sandy or muddy flats. Egg capsules are 5.2 cm long and 3.1 cm wide.

Psammobatis normani McEachran, 1983

Order: Rajiformes

Family: Arhynchobatidae

Common name: Shortfin sandskate

Distribution: Southeast Pacific and Southwest Atlantic

Habitat: Demersal

Description: It is an oviparous species. Eggs have hornlike projections on the shell.

Bathyraja brachyurops (Fowler, 1910)

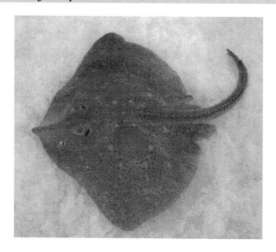

Order: Rajiformes

Family: Arhynchobatidae

Common name: Broadnose skate

Distribution: Southeast Pacific and Southwest Atlantic

Habitat: Demersal; depth range 81–313 m

Description: It has the maximum total length of about 125 cm, which it reaches in about 20 years. Both sexes reach maturity at age 8–10 years. It is oviparous, and the eggs have hornlike projections on the shell.

Bathyraja macloviana (Norman, 1937)

Order: Rajiformes

Family: Arhynchobatidae

Common name: Patagonian skate

Distribution: Southwest Atlantic

Habitat: Demersal

Description: It is a medium-sized skate that has a dark gray to brown body, covered in faint white spots of differing sizes. There are often two large, pale spots in the center of the body, and the entire upper surface is covered with small spines, particularly along the midline of the body. Underparts are white. Body is flattened and fused to the enlarged pectoral fins, forming a broad, flat, diamond-shaped disc. Eyes are positioned on top of the body, just in front of openings known as spiracles. It has a soft, blunt snout and a relatively slender tail, which bears a single dorsal fin. It grows to a maximum size of 77 cm.

Bathyraja scaphiops (Norman, 1937)

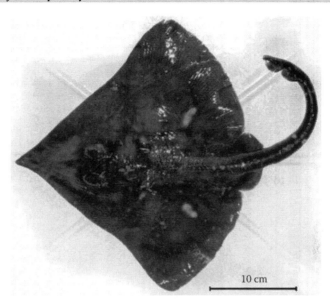

10 cm

Order: Rajiformes

Family: Arhynchobatidae

Common name: Cuphead skate

Distribution: Southwest Atlantic

Habitat: Demersal

Description: Disc of this species is quadrangular to rhomboidal. Mouth is transversed to arched, with numerous teeth. Five pairs of ventral gill slits are present. Tail is very slender, with lateral folds. Usually there are two reduced dorsal fins and a reduced caudal fin. Electric organs are weak and are developed from caudal muscles. Skin is prickly in most species, the prickles often in a row along midline of dorsal.

Nutritional Facts

Proximate Composition (Values Based on Percentage of Wet Mass)

	Moisture (%)	Ash (%)	Lipid (%)	Protein (%)	Energetic Value (kJ/g Wet Mass)
Dipturus chilensis	78.06	2.75	4.42	10.51	4.23
Psammobatis scobina	77.73	2.36	2.05	12.56	3.79
Psammobatis normani	77.90	3.31	1.87	12.13	3.61
Bathyraja brachyurops	75.41	1.92	10.00	11.66	6.72
Bathyraja macloviana	77.64	2.69	3.22	11.68	4.04
Bathyraja scaphiops	78.39	2.65	2.97	11.79	3.97

Source: Data from Eder and Lewis (2005).

Squalus acanthias Linnaeus, 1758

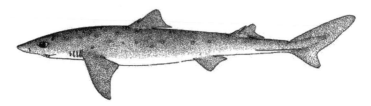

Order: Squaliformes

Family: Squalidae

Common name: Dogfish

Distribution: Western Atlantic

Habitat: Marine; brackish; benthopelagic; oceanodromous; depth range 0–1460 m

Description: This species grows to a common length of 1.6 m and weight of 9.1 kg. This moderately sized species is distinguished by the following set of characteristics: a very slender body, a narrow head, a moderately long snout, a single-lobed anterior nasal flap, and small dorsal fins. Pale caudal fin is poorly demarcated with a whitish margin. A blackish caudal blotch is at the apex of the upper lobe. The dark caudal bar is absent. The dorsal and lateral surfaces of the body are bluish-gray with an irregular array of moderately large white spots, and is whitish ventrally.

Nutritional Facts

Proximate Composition (Values Based on Percentage of Wet Mass)

Moisture (%)	Ash (%)	Lipid (%)	Protein (%)	Energetic Value (kJ/g Wet Mass)
72.79	2.31	8.55	12.79	6.42

Source: Data from Eder and Lewis (2005).

Vitamins (per 200 g)

A	420 µg
D	2 µg
E	4.4 mg
B_1	0.08 mg
B_2	0.16 mg
Niacin	2 mg
B_6	0.66 mg
B_{12}	3.4 µg
Folate	4 µg
B_5	1.46 mg

Minerals (mg/200 g)

Na	K	Ca	Mg	P	Fe	Zn	Cu	Mn
200	900	12	38	400	2	0.6	0.08	0.02

Source: Data from CalorieSlism (2014).

Discopyge tschudii Heckel, 1846

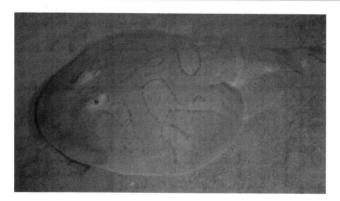

Order: Torpediniformes

Family: Narcinidae

Common name: Apron ray

Distribution: Southeast Pacific

Habitat: Demersal; depth range 5–165 m

Description: It has a circular pectoral fin disc, two dorsal fins, and a stout tail with lateral folds. Its distinguishing feature is its pelvic fins, which are merged beneath the tail to form a continuous disc. The maximum length reported for the apron ray is 54 cm.

10
TURTLES

Phylum: Chordata

Class: Reptilia

Order: Testudines(=Chelonii)

Family: Cheloniidae

Common name: Green sea turtle

Distribution: Throughout tropical and subtropical seas around the world

Habitat: Shallow lagoons

Description: This sea turtle's dorso-ventrally flattened body is covered by a large, teardrop-shaped carapace. It has a pair of large, paddle-like flippers. It is usually lightly colored, although in the eastern Pacific populations, parts of the carapace may be almost black.

Nutritional Facts

Proximate Composition and Fatty Acids (per 28 g)

Protein	Water	Ash	Fat	Omega 3	Omega 6	Cholesterol
5.5 g	22 g	0.3 g	0.1 g	23.8 mg	9.2 mg	14 mg

Vitamins (per 28 g)

A	E	Niacin	Folate	B$_{12}$	Choline
28.0 IU	0.1 mg	0.3 mg	4.2 µg	0.3 µg	18.2 mg

Minerals (mg per 28 g)

Ca	Fe	Mg	P	K	Na	Zn	Cu	Se
33	0.4	5.6	50.4	64.4	19.0	0.3	0.1	4.7*

*µg.

Source: Data from United States Department of Agriculture, National Nutrient Database for Standard Reference, Release 22, 2008.

II
MAMMALS

Eumetopias jubatus (Schreber, 1776)

Phylum: Chordata

Class: Mammalia

Order: Carnivora

Family: Otariidae

Common name: Steller's sea lion

Distribution: Northern Pacific

Habitat: Cool waters, hauling out on beaches and rocky coastline

Description: The Steller's sea lion is the largest of the eared seals. The impressive adult males are two and a half times the size of the females. They have large necks, and shoulders covered with a mane of long, coarse hair. Both sexes are a light brown, whereas pups are originally black, molting to the adult coat after 3–4 months. Length of male and female is 3 m and 2.3 m, respectively. The corresponding values for weight are 700 kg and 300 kg.

Balaena mysticetus Linnaeus, 1758

Phylum: Chordata

Class: Mammalia

Order: Cetartiodactyla

Family: Balaenidae

Common name: Bowhead whale

Distribution: Throughout the high latitudes of the northern hemisphere seas

Habitat: Close to the edge of the Arctic ice shelf

Description: The bowhead whale has a bow-shaped mouth and is black in color with a whitish chin patch broken by a "necklace" of black spots. This whale lacks a dorsal fin and is usually visible as two bumps from above water, which correspond to the head and the back. The blow (or spout) is produced from the two widely spaced blowholes, and it is a bushy V shape reaching 7 m in height. The baleen is dark gray to black in color. Females are larger than males. Maximum length and weight of this species are 20 m and 80 tonnes, respectively.

Erignathus barbatus (Erxleben, 1777)

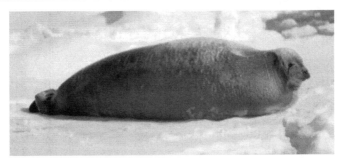

Phylum: Chordata

Class: Mammalia

Order: Carnivora

Family: Phocidae

Common name: Bearded seal (Oogruk)

Distribution: Arctic waters of both the Atlantic and Pacific Oceans

Habitat: Maintains a close association with ice throughout its life

Description: It is typically light to dark gray, with a darker streak running down

262

the middle of the back, and a reddish-brown face and fore flippers. However, coloration is highly variable, and some individuals may be differing shades of brown. This species has small ear holes and short, backward-pointing hind flippers. Maximum length and weight of adults are 2.6 m and 360 kg, respectively.

Delphinapterus leucas (Pallas, 1776)

Phylum: Chordata

Class: Mammalia

Order: Cetartiodactyla

Family: Monodontidae

Common name: Beluga whale

Distribution: Arctic waters

Habitat: Cold arctic waters, usually near the ice edge

Description: The stocky body of this species ends in a particularly small head, and adults develop their striking white coloring as they mature. Beluga whales lack a dorsal fin, but there is a ridge of toughened skin along the back that tends to be more pronounced in mature males. Maximum length of male and female is 5.5 m and 4.1 m, respectively. The weight of this species may be between 500 kg and 1500 kg.

Nutritional Facts (per 28 g)

	SSL	BHW	BS	BW
Proximate composition				
Protein	6.2 g	3.5 g	0.2 g	—
Total carbohydrate	1.6 g	—	—	—
Fats and fatty acids				
Total fat	4.1 g	12.9 g	27.9 g	0.1 g
Saturated fat	0.8 g	1.8 g	3.1 g	—
Monounsaturated	0.9 g	7.9 g	13.2 g	0.1 g
Polyunsaturated	0.8 g	2.2 g	9.2 g	—

Continued

Nutritional Facts (per 28 g)

	SSL	BHW	BS	BW
Vitamins				
A	—	210 IU	9.2 IU	96.2 IU
D	—	—	8.14 IU	—
Riboflavin	—	—	—	0.1 mg
Niacin	—	0.2 mg	—	1.5 mg
Folate	—	—	0.6 cg	1.1 µg
B12	—	—	—	0.7 µg
Pantothenic acid	—	—	—	0.2 mg
Choline	—	—	—	9.8 mg
Betaine	—	—	—	3.9 mg
Minerals				
Ca	1.7 mg	1.4 mg	—	2.0 mg
Fe	2.7 mg	—	—	7.3 mg
Mg	5.3 mg	—	—	6.2 mg
P	59.9 mg	24.4 mg	—	66.9 mg
K	69.9 mg	—	—	79.2 mg
Na	22.4 mg	—	—	21.8 mg
Zn	1.2 mg	—	—	0.8 mg
Se	33.3 µg	—	0.9 µg	10.2 µg
Sterols				
Cholesterol	10.5 mg	15.1 mg	14.6 mg	22.4 mg
Others				
Water	15.9 g	11.2 g	—	20.3 g
Ash	0.3 g	—	—	0.2 g

SSL: *Eumetopias jubatus*; BHW: *Balaena mysticetus*; BS: *Erignathus barbatus*; BW: *Delphinapterus leucas*.
Source: Data from USDA, Release 21.

Odobenus rosmarus (Linnaeus, 1758)

Phylum: Chordata

Class: Mammalia

Order: Carnivora

Family: Odobenidae

Common name: Walrus

Distribution: Discontinuous circum-polar Arctic and sub-Arctic.

Habitat: Shallow waters above the continental shelves; ice floes and beaches on islands or remote stretches of mainland coastlines

Description: Walrus are characterized by their large tusks, which are well developed in both males and females. Tusks are used for interspecific aggression, for defense against predators (polar bears and killer whales), and as an aid for hauling out on ice. Males reach about 3.6 m in length and weigh 880–1557 kg.

Nutritional Facts

Proximate Composition (g/100g)

Protein	Fat	Water	Ash
19.2	13.6	65.10	2.10

Fatty Acids (g/100g)

Saturated fatty acids	2.57
MUFA	8.42
PUFA	2.6
Octadecadienoic acid	0.1
Octacatrienoic acid	0.05
Cholesterol	80*

*mg.
MUFA: monounsaturated fatty acids;
PUFA: polyunsaturated fatty acids.

Vitamins (mg/100g)

Niacin	Riboflavin	Thiamin
4.8	0.24	0.18

Minerals (mg/100g)

Ca	Fe	P
10.0	9.4	122.0

Source: Data from USDA, Release 26.

REFERENCES

Abdul, M., M. Siddique, M. Aktar, and M. A. M. Khatib, 2013. Proximate chemical composition and amino acid profile of two red seaweeds (*Hypnea pannosa* and *Hypnea musciformis*) collected from St. Martin's Island, Bangladesh. *Journal of Fisheries Sciences*, 7: 178–186.

Abdullah, N. M., 2013. Fatty acids profiles of red seaweed, *Gracilaria manilaensis*. *The Experiment*, 11: 726–732.

Abdul-Sahib, I. M., and S. G. Ajeel, 2005. Biochemical constituents and nutritional values for the males and females of the commercial penaeid shrimp *Metapenaeus affinis* (H. Milne–Edwards). *Journal of Basrah Researches (Sciences)*, 31: 35–40.

Ahmad, F. M., R. Sulaiman, W. Saimon, C. F. Yee, and P. Matanjun, 2012. Proximate compositions and total phenolic contents of selected edible seaweed from Semporna, Sabah, Malaysia. *Borneo Science (The Journal of Science and Technology)*, 31: 74–83.

Akköz, C., D. Arslan, A. Ünver, M. M. Özcan, and B. Yilmaz, 2011. Chemical composition, total phenolic and mineral contents of *Enteromorpha intestinalis* Kütz and *Cladophora glomerata* Kütz. Seaweeds. *Journal of Food Biochemistry*, 35: 513–523.

Albert, G., J. Taconi, and M. Metian, 2013. Fish matters: Importance of aquatic foods in human nutrition and global food supply. *Reviews in Fisheries Science*, 21(1): 22–38.

Ambreen, A., K. Hira, A. Tariq, A. Ruqqia, V. Sultana, and J. Ara, 2012. Evaluation of biochemical component and antimicrobial activity of some seaweeds occurring at the Karachi coast. *Pakistan Journal of Botany*, 44: 1799–1803.

Ayas, D., and Y. Özoğul, 2011. The chemical composition of sexually mature blue swimmer crab (*Portunus pelagicus*, Linnaeus 1758) in the Mersin Bay. *Journal of Fisheries Sciences*, 5: 308–316.

Azad Shah, A. K. M., H. Kurihara, and K. Takahashi, 2013. Seasonal variations of lipid content and composition in starfish *Asterias amurensis* Lütken. *Eurasian Chemico-Technological Journal*, 15: 45–50.

Banerjee, S., W. E. Hew, H. Khatoon, M. Shariff, and F. Yusoff, 2011. Growth and proximate composition of tropical marine *Chaetoceros calcitrans* and *Nannochloropsis oculata* cultured outdoors and under laboratory conditions. *African Journal of Biotechnology*, 10: 1375–1383.

Barrento, S., A. Marques, B. Texeira, P. Vaz-Pires, and M.L. Nunes, 2009. Nutritional quality of the edible tissues of European lobster *Homarus gammarus* and American lobster *Homarus americanus*. *Journal of Agricultural Food Chemistry*, 57: 3645–3652.

Benjama, O., and P. Masniyom, 2011. Nutritional composition and physicochemical properties of two green seaweeds (*Ulva pertusa* and *Ulva intestinalis*) from the Pattani Bay in Southern Thailand. *Songklanakarin Journal of Science and Technology*, 33: 575–583.

Bordbar, S., F. Anwar, and N. Saari, 2011. High-value components and bioactives from sea cucumbers for functional foods—A review. *Marine Drugs*, 9: 1761–1805.

Bykov, V. P., 1983. *Marine Fishes: Chemical Composition and Processing Properties*. New Delhi: Amerind.

Chen, Y., T. Chen, T. Chiou, and D. Hwang, 2013. Seasonal variation on general composition, free amino acids and fatty acids in the gonad of Taiwan's sea urchin *Tripneustes gratilla*. *Journal of Marine Science and Technology*, 21: 723–732.

Chrishanthi, E. K. I., and M. V. E. Attygalle, 2011. Fat contents and fatty acid profiles of Indian scad. *Vidyodaya Journal of Science*, 16. (Abstract.)

Department of Nutrition, 2009. Mørkhøj Bygade 19–DK-2860. Søborg, Denmark: Department of Nutrition.

Dhaneesh, K. V., K. M. Noushad, and T. A. Kumar, 2012. Nutritional evaluation of commercially important fish species of Lakshadweep Archipelago, India. DOI:10.1371/journal.pone.0045439

Eder, E. B., and M. N. Lewis, 2005. Proximate composition and energetic value of demersal and pelagic prey species from the SW Atlantic Ocean. *Marine Ecology Progress Series*, 291: 43–52.

Fayaz, M., K. K. Namitha, K. N. Murthy, M. M. Swamy, R. Sarada, S. Khanam, P. V. Subbarao, and G. A. Ravishankar, 2005. Chemical composition, iron bioavailability, and antioxidant activity of *Kappaphycus alvarezzi* (Doty). *Journal of Agricultural and Food Chemistry*, 53: 792–797.

Ghada F. E., and A. El-Sikaily, 2013. Chemical composition of some seaweed from Mediterranean Sea coast, Egypt. *Environmental Monitoring and Assessment*, 185: 6089–6099.

Goecke, F., M. Escobar, and G. Collantes, 2012. Chemical composition of *Padina fernandeziana* (Phaeophyceae, Dictyotales) from Juan Fernandez Archipelago, Chile. *Latin American Journal of Aquatic Research*, 3: 95–104.

González, M. J., J. M. Gallardo, P. Brickle, and I. Medina, 2007. Nutritional composition and safety of *Patagonotothen ramsayi*, a discard species from Patagonian Shelf. *International Journal of Food Science & Technology*, 42: 1240–1248.

Gunalan B., S. N. Tabitha, P. Soundarapandian, and T. Anand, 2013. Nutritive value of cultured white leg shrimp *Litopenaeus vannamei*. *International Journal of Fisheries and Aquaculture*, 5: 166–171.

Holland, B., J. Brown, and D. H. Buss, 1993. Fish and fish products. Third Supplement to 5th edition of McCance and Widdowson's *The composition of foods*. Cambridge: Royal Society of Chemistry.

Hossain, M. A., 2011. Fish as source of n-3 polyunsaturated fatty acids (PUFAs), Which one is better—Farmed or wild? *Advance Journal of Food Science and Technology*, 3: 455–466.

Hwang, E., K. Ki, and H. Chung, 2013. Proximate composition, amino acids, mineral and heavy metal content of dried layer. *Preventive Nutrition and Food Science*, 18: 139–144.

Jayasankar, R., 1993. Seasonal variation in biochemical constituents of *Sargassum wightii* (Grevillie) with reference to yield in alginic acid content. *Seaweed Research and Utilization*, 16: 13–16.

Khairy, H. M., and S. M. El-Shafay, 2013. Seasonal variations in the biochemical composition of some common seaweed species from the coast of Abu Qir Bay, Alexandria, Egypt. *Oceanologia*, 55: 435–452.

Kolb, N., L. Vallorani, N. Milanovi, and V. Stocchi, 2004. Evaluation of marine algae wakame (*Undaria pinnatifida*) and kombu (*Laminaria digitata* japonica) as food supplements. *Food Technology and Biotechnology*, 42: 57–61.

Krzynowek, J., and J. Murphy, 1987. Proximate Composition, Energy, Fatty Acid, Sodium, and Cholesterol

Content of Finfish, Shellfish, and Their Products. NOAA Technical Report NMFS 55, p. 53.

Lee, M. H., 2012. Comparison on proximate composition and nutritional profile of red and black sea cucumbers (*Aposticophus japonicus*) from Ulleungdo (island) and Dovado (island), Korea. *Food Science and Biotechnology*, 21: 1285–1291.

Lua, D., M. Zhanga, S. Wangc, J. Cai, X. Zhoud, and C. Zhud, 2010. Nutritional characterization and changes in quality of *Salicornia bigelovii* Torr. during storage. *LWT–Food Science and Technology*, 43: 519.

MacArtain, P., C. I. R. Gill, M. Brooks, R. Campbell, and I. R. Rowland, 2007. Nutritional value of edible seaweeds. *Nutrition Reviews*, 65: 535–543.

Mathana, P., S. Thiravia Raj, C. Radha Krishnan Nair, and T. Selvamohan, 2012. Seasonal changes in the biochemical composition of four different tissues of red spotted emperor *Lethrinus Lentjan* (Family: Lethrinidae). *Annals of Biological Research*, 3 (11): 5078–5082.

Matunjan, P., S. Mohamed, N. M. Mustapha, and K. Mukammad, 2009. Nutrient content of tropical edible seaweeds, *Eucheuma cottonii*, *Caulerpa lentillifera* and *Sargassum polycystem*. *Journal of Applied Phycology*, 21: 75–80.

Moini, S., Zh. Khoshkhoo, and R. Hemati Matin, 2012. The fatty acids profile in mackerel (*Scomberomorus guttatus*) and its shelf life in cold storage at −18°C. *Global Veterinaria*, 8: 665–669.

Nasopoulou, C., H. C. Karantonis, and I. Zabetakis, 2011. Nutritional value of gillhead sea bream and sea bass. *Dynamic Biochemistry, Process Biotechnology and Molecular Biology*, 5: 32–40.

Nisa, K., and R. Sultana, 2010. Variation in the proximate composition of shrimp, *Fenneropenaeus penicillatus*

at different stages of maturity. *American-Eurasian Journal of Scientific Research*, 5: 277–282.

Norziah, M. H., and C. Y. Ching, 2000. Nutritional composition of edible seaweed *Gracilaria changgi*. *Food Chemistry*, 68: 69–76.

Nurnadia, A. A., A. Azrina, and I. Amin, 2011. Proximate composition and energetic value of selected marine fish and shellfish from the west coast of peninsular Malaysia. *International Food Research Journal*, 18: 137–148.

Nurnadia, A. A., A. Azrina, I. Amin, A. S. M. Yunus, and M. I. H. Effendi, 2013. Mineral contents of selected marine fish and shellfish from the west coast of peninsular Malaysia. *International Food Research Journal*, 20: 431–437.

Öksüz, A., A. Ozyilmaz, M. Aktas, G. Gercek, and J. Motte, 2009. A comparative study on proximate, mineral and fatty acid compositions of deep seawater rose shrimp (*Parapenaeus longirostris*, Lucas 1846) and golden shrimp (*Plesionika martia*, A. Milne-Edwards). *Journal of Animal and Veterinary Advances*, 8: 183–189.

Öksüz, A., Iskenderun-Turkey, A. Özyılmaz, and H. Sevimli, 2010. Element compositions, fatty acid profiles, and proximate compositions of marbled spinefoot (*Siganus rivulatus*, Forsskäl, 1775) and dusky spinefoot (*Siganus luridus*, Ruppell, 1878). *Journal of Fisheries Sciences*, 4: 177–183.

Osman, H., A. R. Suriah, and E. C. Law, 2001. Fatty acid composition and cholesterol content of selected marine fish in Malaysian waters. *Food Chemistry*, 73: 55–60.

Palanikumar, M., A. Ruba Annathai, R. Jeya Shakila, and S. A. Shanmugam, 2014. Proximate and major mineral composition of 23 medium sized marine fin fishes landed in the Thoothukudi Coast of India. *Journal of Nutrition & Food Sciences*, 4(1): 1000259.

Pathirana, L., C., Shahidi, F. Whittick, and Alan, 2002. Comparison of nutrient composition of gonads and coelomic fluid of green sea urchin *Strongylocentrotus droebachiensis*. *Journal of Shellfish Research*, 21: 861–870.

Pennywhite, 2007. Fountain of youth equals, *Porphyra* sp. (It wouldn't hurt). *The Science Creative Quarterly*.

Periyasamy, N., M. Srinivasan, K. Devanathan, and S. Balakrishnan, 2011. Nutritional value of gastropod *Babylonia spirata* (Linnaeus, 1758) from Thazhanguda, Southeast coast of India. *Asian Pacific Journal of Tropical Biomedicine*, S249–S252.

Polat, S., and Y. Ozogul, 2013. Seasonal proximate and fatty acid variations of some seaweeds from the northeastern Mediterranean coast. *Oceanologia*, 55: 2013.

Puga-López, D., J. T. Ponce-Palafox, G. Barba-Quintero, M. Rosalía Torres-Herrera, E. Romero-Beltrán, and J. L. Arredondo Figueroa, 2013a. Proximate composition and microbiological muscle properties, in two species shrimps of the Pacific tropical coast. *Journal of Agricultural Science*, (Appl. Volume) 2: 151–154.

Puga-López, D., J. T. Ponce-Palafox, G. Barba-Quintero, M. Rosalía Torres-Herrera, E. Romero-Beltrán, J. L. Arredondo Figueroa, and M. García-ulloa Gomez, 2013b. Sensory analysis of farmed and wild harvested white shrimp *Litopenaeus vannamei* (Boone, 1931) tissues. *Current Research Journal of Biological Sciences*, 5: 130–135.

Pushparajan, N., P. Soundarapandian, and D. Varadharajan, 2012. Proximate composition of fresh and prepared meats stored in tin free steel cans. *Journal of Marine Science: Research and Development*, 2: 4.

Rameshkumar, S., C. M. Ramakritinan, and M. Yokeshbabu, 2013. Proximate composition of some selected seaweeds from Palk Bay and Gulf of Mannar, Tamil Nadu, India. *Asian Journal of Biomedical and Pharmaceutical Sciences*, 3: 1–5.

Ratana-arporn, P., and A. Chirapart, 2006. Nutritional evaluation of tropical green seaweeds *Caulerpa lentillifera* and *Ulva reticulata*. *Kasetsart Journal (Natural Science)*, 40 (Suppl.): 75–83.

Rizvi, M. A., and M. Shameel, 2010. Elemental composition in various thallus parts of three brown seaweeds from Karachi coast. *Pakistan Journal of Botany*, 42: 4177–4181.

Rodriguez, A. P., T. P. Mawhinney, D. Ricque-Marie, and L. E. Cruz-Suarez, 2011. Chemical composition of cultivated seaweed *Ulva clathrata*. *Food Chemistry*, 129: 491–498.

Salam, H. A. A., 2013. Evaluation of nutritional quality of commercially cultured Indian white shrimp *Penaeus indicus*. *International Journal of Nutrition and Food Sciences*, 2: 160–166.

Seenivasan, R., M. Rekha, H. Indu, and S. Geetha, 2012. Antibacterial activity and phytochemical analysis of selected seaweeds from Mandapam Coast, India. *Journal of Applied Pharmaceutical Science*, 2: 159–169.

Shanmugam, A., and C. Palpandi, 2008. Biochemical composition and fatty acid profile of the green alga *Ulva reticulata*. *Asian Journal of Biochemistry*. 3: 26.

Siddique, M. A. M., M. Aktar, and M. A. M. Khatib, 2013a. Proximate chemical composition and amino acid profile of two red seaweeds (*Hypnea pannosa* and *Hypnea musciformis*) collected from St. Martin's Island, Bangladesh. *Journal of Fisheries Sciences*, 7: 178–186.

Siddique, M. A. M., M. S. K. Khan, and M. K. A Bhuiyan, 2013b. Nutritional composition and amino acid profile of a sub-tropical red seaweed *Gelidium pusillum* collected from

St. Martin's Island, Bangladesh. *International Food Research Journal,* 20(5): 2287–2292.

Siddique, M. A. M., P. Mojumder, and H. Zamal, 2012. Proximate composition of three commercially available marine dry fishes (*Harpodon nehereus, Johnius dussumieri,* and *Lepturacanthus savala*). *American Journal of Food Technology,* 7: 429–436.

Siong, T. E., S. M. Shahid, R. Kuladevan, Y. S. Ing, and K. S. Choo, 1987. Nutrient composition of Malaysian marine fishes. *Asian Food Journal,* 3 (2).

Siron, R., G. Giusti, and B. Berland, 1989. Changes in the fatty acid composition of *Phaeodactylum tricornutum* and *Dunaliella ertiolecta* during growth and under phosphorus deficiency. *Marine Ecology Progress (Series),* 5: 95–100.

Sridharan M. C., and R. Dhamotharan, 2012. Amino acids and fatty acids in *Turbinaria conoides. Journal of Chemical and Pharmaceutical Research,* 4: 5093–5097.

Stansby, M. E., 1987. Nutritional properties of recreationally caught marine fishes. *Marine Fisheries Review,* 49(2): 118–121.

Sudhakar, M., K. Manivannan, and P. Soundrapandian, 2009. Nutritive value of hard and soft shell crabs of *Portunus sanguinolentus* (Herbst). *International Journal of Animal and Veterinary Advances,* 1: 44–48.

Suseno, S. H., R. R. T. F. Permata, S. Hayati, R. Nugraha, and Saraswati, 2013. Fatty acid composition and steroid content of selected deep sea fishes (escolar (*Lepidocibium plavobrunneum*) and deep sea lobster (*Linuparus somniosus*)) from Southern Java Ocean–Indonesia. Paper presented at the 2nd International Conference on Ecological, Environmental and Bio-Sciences (ICEEBS'2013), December 20–21, Bali, Indonesia.

Tabarsa, M., M. Rezaei, Z. Ramezanpour, and J. R. Waaland, 2012. Chemical compositions of the marine algae *Gracilaria salicornia* (Rhodophyta) and *Ulva lactuca* (Chlorophyta) as a potential food source. *Journal of the Science of Food and Agriculture,* 92: 2500–2506.

Tamilselvi, M., V. Sivakumar, H. A. J. Ali, and R. D. Thilaga, 2010. Preparation of pickle from *Herdmania pallida,* simple ascidian. *World Journal of Dairy & Food Sciences,* 5: 88–92.

Udo, P. J., and V. N. Arazu, 2012. Nutritive value of *Ethmalosa fimbraita* (Clupeidae), *Mugil cephalus,* (Mugilidae) and *Cynoglosuss senegalensis* (Cynoglossidae) of the Cross River Estuary, Nigeria, West Africa. *Pakistan Journal of Nutrition,* 11 (7): 526–530.

U.S. Department of Agriculture, 2008. National Nutrient Database for Standard Reference 1, Release 22.

USDA (U.S. Department of Agriculture), 1987. Human Nutrition Information Service Handbook Number 8-15, 1987.

U.S. Department of Agriculture, 2013. National Nutrient Database for Standard Reference- Release 26.

Wan Rosli, W. I., A. J. Rohana, S. H. Gan, H. N. Fadzlina, H. Rosliza, H. H. M. S. Nazri, M. Ismail, S. Bahri, W. B. W. Mohamad, and K. Imran, 2012. Fat content and EPA and DHA levels of selected marine, freshwater fish and shellfish species from the east coast of peninsular Malaysia. *International Food Research Journal,* 19: 815–821.

Xiguang, L., Y. Huahua, Z. Zengqin, L. Zhien, and L. Pengcheng, 2004. Study on the fatty acid composition of jellyfish *Rhopilema esculentum. Chinese Journal of Analytical Chemistry,* 32: 1635–1638.

INDEX

A

Abalistes stellaris, 239
Abrupt wedge shell, 103
Acanthistius brasilianus, 200
Acanthopora spicifera, 55
Acaudina molpadioides, 131
Acetes japonicas, 80
Actinopyga echinites, 135
Adipose eyelid, 158, 159, 237
African spear lobster, 93
Agar-agar Lesong, 29
Akiami paste shrimp, 81
Alaska pollock, 230
Albacore, 196
American eel, 233
American jackknife clam, 101
American lobster, 92
Amphidromous, 141, 154,
    156, 163, 167, 186,
    192, 205, 211–213,
    216, 217, 237–240,
    246, 250
Amphipods, 224
Anadontosoma chacunda, 148
Anadromous, 143, 145, 149,
    177, 245
Anchovy, 149
Anguilla rostrate, 233
Antarctic armless
    flounder, 215
Apostichopus japonicas, 132
Apron ray, 257
Apron-ribbon vegetable, 43
Arctica islandica, 102
Areolate grouper, 201
Argentine hake, 232
Argentine seabass, 200
Argentine short fin squid, 121
Argentinian sandperch, 183
Arrowtooth flounder, 220
Asakusa nori, 67
Ascophyllum nodosum, 23
Asian kelp, 43
Asian tiger shrimp, 79
Asiatic hard clam, 103
Astaxanthin, 8
Asterias amurensis, 127
Atheresthes stomias, 220

Atlantic cod, 227
Atlantic herring, 2, 146
Atlantic mackerel, 2
Atlantic pomfret, 150
Atractoscion nobilis, 188
Atrial siphon, 137
Austrophycis marginata, 233

B

Babylonia spirata, 113
Balaena mysticetus, 262
Baleen, 262
Banded rattail, 230
Barnacles, 130
Barramudi, 165
Bassanago albescens, 234
Bathyraja brachyurops, 253, 255
Bathyraja macloviana, 254, 255
Bathyraja scaphiops, 255
Bearded mussel, 100
Bearded seal, 262
Beluga whale, 263
Benthopelagic, 147, 153,
    160, 161, 182, 200,
    210–213, 226,
    230–233, 237, 247,
    248, 256
Bigeye snapper, 169
Bilateral symmetry, 135
Bioluminescent, 233
Black abalone, 115
Black clam, 102
Black lip abalone, 117
Black pomfret, 153
Black teatfish, 133
Blackfoot limpet, 119
Blackfoot opihi, 119
Blood-spotted abalone, 115
Blood-spotted swimming
    crab, 84
Blubber weed, 48
Blue crab, 83, 86
Blue hypnea, 49
Blue mackerel, 197
Blue mussel, 99
Blue shrimp, 80
Bluefin tuna, 194
Bluefish, 186

Bluespot mullet, 236
Blue-striped snapper, 168
Bombay duck, 247
Bonga shad, 147
Bowhead whale, 262
Brain food, 4
Brama brama, 150
Branchial siphon, 137
Breeding tubercles, 245
Bright-green nori, 17
Broadnose skate, 254
Bullacta exarata, 114
Butter clam, 105
Buttonweed, 24
Byssal threads, 100

C

Cactus algae, 15
Caesio erythrogaster, 151
California red abalone, 117
Callinectes sapidus, 85
Callorhynchus callorhynchus,
    247
Calpomenia sinuosa, 30
Cancer magister, 86
Candlefish, 245
Cannonball jelly, 72
Canthaxanthin, 8
Capelin, 218, 245
Carangoides malabaricus, 156
Carangoides orthogrammus, 157
Caranx djeddaba, 159
Caudal peduncle, 153, 154,
    192, 199, 211, 233,
    235, 240
Caulerpa lentillifera, 12
Caulerpa racemosa, 13
Cellana exarata, 119
Channel bull blenny, 150
Checkerboard, 170
Checkered carpet shell, 106
Checkered snapper, 169
Cheilinus undulatus, 164
Chelonia mydas, 259
Chilean sea bass, 181
Chinese oyster, 96
Chinese scallop, 110
Chinese silver pomfret, 211

Chionoecetes bairdi, 89
Chionoecetes opilio, 88
Chirocentrus dorab, 141
Chlamys farreri, 110
Chnoospora minima, 31
Chondrus crispus, 45
Chorinemus lysan, 158
Christmas wrasse, 163
Chub mackerel, 192
Cladophora glomerata, 16
Clupea harengus, 146
Cockfish, 247
Cockscomb, 131
Codium iyengarii, 12
Codium tomentosum, 11
Coelorhynchus fasciatus, 230
Coenocytes, 14
Coho salmon, 224
Collector urchin, 128
Common forked tongue, 32
Common orient clam, 104
Congiopodus peruvianus, 240
Coontail, 31
Copepods, 137
Coral cod, 202
Cottoperca gobio, 149
Cottunculus granulosus, 241
Crassostrea gigas, 97
Crassostrea gryphoides, 95
Crassostrea madrasensis, 96
Crassostrea rivularis, 96
Cross-cut carpet shell, 106
Crozier weeds, 50
Cuphead skate, 255
Cuvierian tubules, 134
Cynoglossus arel, 216
Cynoglossus lingua, 217
Cynoglossus senegalensis, 215
Cynoscion nebulosus, 189
Cynoscion nobilis, 188
Cynoscion regalis, 189

D

Dasya rigidula, 56
Dasyatis zugei, 250
Date mussel, 100
Date shell, 100
Decapterus russelli, 160
Deep seawater rose shrimp, 77
Deepwater redfish, 135
Delagoa threadfin bream, 179
Delphinapterus leucas, 263, 264
Devil's thumbprint, 228
Diatoms, 89, 137
Dicentrarchus labrax, 178

Dictyota dichotoma var.
    intricata, 32
Dictyota indica, 33
Dillisk, 3
Diplodus sargus, 207
Dipturus chilensis, 252, 255
Discopyge tschudii, 247, 257
Dissostichus eleginoides, 181
Dogfish, 248, 256
Donax trunculus, 102
Doublespotted queen fish, 158
Drepane punctata, 162
Duckbill, 243
Dulse, 3
Dungeness crab, 86
Dussumier's halfbeak, 235
Dussumier's ponyfish, 166
Dussumieria acuta, 142
Dussumieria hasselti, 146
Dwarf codling, 233
Dwarf gelidium, 57
Dwarf glasswort, 70
Dwarf saltwort, 70

E

Eelgrass beds, 86
Eelpout, 214
Eisenia arborea, 40
Electric organs, 255
Elephantfish, 247
Eleutheronema tetradactylum,
    185
Ensis directus, 101
Enteroctopus dofleini, 124
Enteromorpha, 15
Enteromorpha clathrata, 17
Epinephelus areolatus, 201
Epinephelus sexfasciatus, 201
Epinephelus tauvina, 202
Epipodium, 116
Erignathus barbatus, 262, 264
Esox lucius, 243
Estuarine oyster, 95
Ethmalosa fimbriata, 147
Eucheuma alvarezii, 52
Eucheuma cottonii, 46
Eucheuma denticulatum, 47
Eucheuma spinosum, 47
Eucheuma striatum, 54
Eulachon, 245
Eumetopias jubatus, 261, 264
Euphausiids, 77
European anchovy, 2
European edible abalone, 116
European lobster, 91

European sea bass, 178
Euthynnus affinis, 196
Ezo giant scallop, 111

F

Fan mussel, 100
Fan-leaf seaweed, 34
Fathead, 241
Fenneropenaeus
    penicillatus, 76
Filiform sea moss, 62
Five-lined snapper, 168
Flippers, 259, 263
Flower crab, 83
Four-finger threadfin, 186
Fringe scale sardinella, 143

G

Gadus macrocephalus, 227
Gadus morhua, 227
Galeocerdo cuvieri, 248
Gazza achlamys, 166
Gelidium pusillum, 57
Genypterus blacodes, 244
Geoduck clam, 109
Geryon quinquedens, 88
Giant ezo scallop, 111
Giant horse mussel, 100
Giant mud crab, 85
Giant oyster, 95
Giant sea catfish, 239
Giant seaperch, 165
Giant tiger prawn, 79
Gillhead sea bream, 208
Gizzard shad, 148
Golden cuttlefish, 123
Golden shrimp, 77
Golden snapper, 174
Gorgonian, 174
Gracilaria changgi, 60
Gracilaria compressa, 61
Gracilaria cornea, 60
Gracilaria manilaensis, 62
Gracilaria salicornia, 63
Gracilaria verrucosa, 64
Gray eel-catfish, 238
Gray mullet, 237
Greasy grouper, 203
Great amberjack, 153
Greater lizardfish, 246
Greater pony fish, 167
Green ormer snail, 116
Green sea turtle, 259
Green sea urchin, 130

Green tidal-flat mats, 15
Guso, 46
Gymnura poecilura, 251

H

Haddock, 218, 228
Hairy basket weed, 48
Hairy conger, 234
Halibut, 218, 219, 222
Halimeda tuna, 14
Haliotis cracherodii, 115
Haliotis discus hannai, 118
Haliotis gigantea, 117
Haliotis rubra, 117
Haliotis rufescens, 116
Haliotis spadicea, 115
Haliotis tuberculata, 116
Hallow-green nori, 25
Halocynthia roretz, 137
Hard clam, 104, 107
Hardtail scad, 155
Harpodon nehereus, 246
Hawkfish, 162
Hectocotylus, 121
Hen clam, 108
Herdmania pallida, 138
Hilsa (Clupea) macrura, 145
Himanthalia elongata, 24
Hippoglossus
    hippoglossus, 218
Hippoglossus slenolepis, 219
Holothuria (Lessonothuria)
    insignis, 134
Holothuria (Lessonothuria)
    multipilula, 135
Holothuria (Metriatyla)
    scabra, 133
Holothuria (Microthele)
    nobilis, 133
Holothuria (Thymiosycia)
    impatiens, 134
Holothuria pardalis, 134
Homarus americanus, 92
Homarus gammarus, 91
Hooded oyster, 98
Hormophysa cuneiformis, 25
Hornwort, 31
Horse mussel, 100
Horsefish, 240
Houndfish, 235
Humpback red snapper, 171
Humphead wrasse, 164
Hypnea charoides, 51
Hypnea japonica, 50
Hypnea musciformis, 50

Hypnea pannosa, 49
Hyporhamphus dussumieri, 235

I

Ilisha elongata, 144
Ilisha melanostoma, 142
Illex argentines, 121
Iluocoetes fimbriatus, 214
Indian backwater oyster, 96
Indian dogfish, 248
Indian halibut, 222
Indian ilisha, 142
Indian mackerel, 193
Indian oil sardine, 143
Indian pellona, 143
Indian shad, 160
Indian white prawn, 82
Indo-Pacific king
    mackerel, 199
Irish moss, 45
Island trevally, 157

J

Jania rubens, 66
Japanese abalone, 118
Japanese amberjack, 161
Japanese baking scallop, 111
Japanese common squid, 120
Japanese disc abalone, 118
Japanese edible jellyfish, 71
Japanese flying squid, 120
Japanese littleneck clam, 104
Japanese oyster, 97
Japanese red algae, 51
Japanese scallop, 110
Japanese spiky sea
    cucumber, 132
Japanese spineless
    cuttlefish, 123
Japanese threadfin bream, 180
Jellyballs, 72
Jinga shrimp, 81
Jinjiang oyster, 96
Johnius (Pseudosciaena)
    soldado, 191
Johnius dussumieri, 191

K

Kappaphycus alvarezii, 52
Kappaphycus cottonii, 46
Kappaphycus striatum, 54
Karalla dussumieri, 165
Killer whales, 265
King mackerel, 194

Knotted wrack, 23
Korean mud snail, 114

L

Ladder wrasse, 163
Laminaria digitata, 41
Laminaria saccharina, 42
Large tuna, 196
Largehead hairtail, 212
Large-scale tongue sole, 216
Lates calcarifer, 164
Laurencia papillosa, 58
Leiognathus dussumieri, 165
Leiognathus equulus, 167
Leiostomus xanthurus, 190
Leopard sea cucumber, 135
Lepturacanthus savala, 213
Lethrinus lentjan, 168
Limey petticoat, 36
Linuparus somniosus, 93
Lithophaga lithophaga, 100
Litopenaeus stylirostris, 80
Litopenaeus vannamei, 78
Long tongue sole, 217
Long-necked clam, 108
Longspine porgy, 207
Longtail butterfly ray, 251
Longtail shad, 145
Longtail southern cod, 182
Lutjanus argentimaculatus,
    170
Lutjanus bohar, 175
Lutjanus decussatus, 169
Lutjanus gibbus, 171
Lutjanus johnii, 173
Lutjanus lutjanus, 169
Lutjanus malabaricus, 174
Lutjanus quinquelineatus, 168
Lutjanus russelli, 172

M

Macruronus magellanicus, 231
Mactra sachalinensis, 107
Malabar blood snapper, 174
Malabar trevalley, 156
Mallotus villosus, 244
Mancopsetta maculate, 215
Mangrove crab, 85
Mangrove red snapper, 170
Manila clam, 104
Mantis shrimps, 239
Marbled spinefoot, 203
Mediterranean mussel, 99
Megalapsis cordyla, 155

Melanogrammus aeglefinnus, 228
Melanophores, 231
Mercenaria mercenaria, 107
Meretrix lusoria, 104
Meretrix meretrix, 103
meriodional symmetry, 135
Merluccius australis, 231
Merluccius hubbsi, 232
Metapenaeus affinis, 81
Micromesistius australis, 226
Mitre squid, 122
Miyagi oyster, 97
Modiolus (Modiolus) modiolus, 100
Molting, 261
Money shell, 105
Monodactylus argenteus, 176
Moolgarda seheli, 236
Moon fish, 176
Morone saxatilis, 177
Mother-of-pearl, 118
Mugil cephalus, 236
Mya arenaria, 108
Mytilus edulis, 98
Mytilus galloprovincialis, 99

N

Nacre, 118
Nannochloropsis gaditana, 8
Nannochloropsis oculata, 7
Narrow-barred Spanish mackerel, 199
Narrowmouthed catshark, 249
Needle fishes, 235
Nemadactylus bergi, 162
Nematodes, 166
Nemipterus bipunctatus, 179
Nemipterus bleekeri, 179
Nemipterus japonicus, 180
Neon flying squid, 121
Netuma thalassina, 239
New Zealand southern blue whiting, 226
Nibea soldado, 191
North Pacific giant octopus, 125
Northern pike, 243
Notophycis marginata, 233

O

Oarweed, 41
Obtuse barracuda, 209

Ocean quahog, 102
Ocean shrimp, 75
Oceanodromous, 143, 147, 150, 157, 161, 170, 174, 178, 182, 186, 188–199, 203, 204, 208, 210, 213, 218, 219, 221, 223, 226, 228, 229, 231, 232, 235, 241, 244, 247, 248, 256
Ocellated octopus, 125
Octopus ocellatus, 125
Odobenus rosmarus, 264
Okhotsk atka mackerel, 241
Ommastrephes bartramii, 121
Oncorhynchus kisutch, 224
Oncorhynchus mykiss, 225
Oncorhynchus nerka, 224
One-finlet scad, 158
Oogruk, 262
Ophiuroids, 171
Ovoviviparous, 250

P

Pacific abalone, 118
Pacific cod, 227
Pacific flying squid, 120
Pacific halibut, 219
Pacific herring, 2
Pacific littleneck clam, 106
Pacific oyster, 97
Pacific pink shrimp, 75
Padina australis, 34
Padina durvillei, 34
Padina fernandeziana, 35
Padina gymnospora, 36
Padina pavonica, 37
Padina tetrastromatica, 38
Padina vickersiae, 38
Painted sweetlip, 184
Pale-edged stingray, 250
Pallial line, 102
Pallial sinus, 103
Palmaria palmate, 3, 65
Pampus argenteus, 210
Pampus chinensis, 211
Pandalus borealis, 75
Pandalus jordani, 75
Panopea generosa, 109
Panulirus interruptus, 90
Paralichthys patagonicus, 217
Paralithodes camtschatica, 87
Parapenaeus longirostris, 77
Parastromateus niger, 153

Parona leatherjacket, 152
Parona signata, 152
Parupeneus bifasciatus, 178
Patagonian flounder, 217
Patagonian grenadier, 231
Patagonian redfish, 242
Patagonian skate, 254
Patagonotothen ramsayi, 182
Patinopecten yessoensis, 110
Peacock's tail, 37
Pecten albicans, 111
Pelagic red crab, 89
Pellona ditchela, 142
Penaeus indicus, 82
Penaeus monodon, 79
Perca fluviatilis, 183
Periostracum, 99
Periwinkles, 130
Phaeodactylum tricornutum, 9
Pink cusk-eel, 244
Pink ling, 244
Pink shrimp, 76
Plaice, 221
Plectorhinchus pictus, 184
Plesionika martia, 77
Pleurogammus azonus, 241
Pleuroncodes planipes, 89
Pleuronectes platessa, 220
Plotosus canius, 238
Plotosus lineatus, 237
Polar bears, 265
Pollachius virens, 229
Pollock, 229, 230
Polychromic shell, 103
Pomadasys hasta, 185
Pomatomus saltatrix, 186
Porphyra haitanensis, 67
Porphyra spp., 3
Porphyra tenera, 67
Porphyra umbilicalis, 68
Portunus pelagicus, 83
Portunus sanguinolentus, 84
Prickly redfish, 131
Prionotus nudigula, 242
Pristipomoides typus, 173
Protandric hermaphrodite, 75
Prototothaca staminea, 105
Protozoans, 137
Psammobatis normani, 253, 255
Psammobatis scobina, 252, 255
Psettodes erumei, 222
Pseudopercis semifasciata, 183
Pterocladia capillacea, 59
Purple laver, 69

Q

Queen conch, 112

R

Rabbitfish, 203
Rainbow sardine, 142
Raspthorn sandskate, 253
Rastrelliger kanagurta, 193
Ray's bream, 150
Razor clam, 101
Red deepsea crab, 89
Red drum, 188
Red dulse, 65
Red king crab, 87
Red ocean squid, 121
Red sea urchin, 130
Red spotted emperor, 168
Red squid, 121
Redbelly yellowtail fusilier, 151
Redtail prawn, 76
Rhizoclonium implexum, 15
Rhopilema esculentum, 71
Ribbed carpet shell, 106
Ribbon sea lettuce, 22
Round herring, 146
Round seagrapes, 12
Russell's snapper, 172

S

Saccharina latissima, 42
Saccostrea cucullata, 97
Salicornia bigelovii, 69
Salilota australis, 232
Salmo gairdneri, 225
Sandfish, 133
Sardinella fimbriata, 143
Sardinella longiceps, 143
Sargassum polycystum, 28
Sargassum tenerrimum, 26
Sargassum variegatum, 26,
Sargassum wightii, 27
Saurida tumbil, 246
Savalai hairtail, 213
Saxidomus giganteus, 105
Scampi, 244
Scaophthalmus maximus, 223
Schools, 166
Schroederichthys bivius, 249
Sciaena dussumieri, 191
Sciaenops ocellatus, 188
Scoliodon laticaudus, 248
Scoliodon sorrokowah, 248
Scolopsis bimaculatus, 180

Scomber australasicus, 197
Scomber japonicus, 192
Scomber scombrus, 197
Scomberomorus cavalla, 193
Scomberomorus commerson, 199
Scomberomorus guttatus, 199
Scylla serrata, 85
Sea balloon, 30
Sea grapes, 13
Sea lettuce, 3, 18, 19, 20
Sea pineapple, 137
Sea potato, 30
Sea salmon, 183
Sea star, 127
Sebastes oculatus, 242
Selar mate, 157
Selaroides leptolepis, 154
Selenizone, 115
Senegalese tonguesole, 215
Sepia esculenta, 123
Sepiella japonica, 123
Seriola lalandi, 153
Seriola quinqueradiata, 160
Seriolella punctata, 161
Sharptooth bass, 173
Shortfin sandskate, 253
Siebold's abalone, 118
Siganus javus, 204
Siganus rivulatus, 203
Sillago maculate, 206
Sillago sihama, 205
Silver grunter, 185
Silver Jewfish, 192
Silver pomfret, 210, 211
Silver warehou, 161
Silver whiting, 205
Sin croaker, 191
Sixbar grouper, 202
Slender shad, 144
Small agar weed, 59
Smalltoothed ponyfish, 166
Smoked yellow stripe
        trevally, 154
Snow crab, 88
Snubnose pomp, 152
Sockeye salmon, 224
Soft-shell clam, 108
Soldier croaker, 192
Solieria robusta, 47
Southern hake, 231
Southern sea palm, 40
Southwest Atlantic
        butterfish, 209
Spanish mackerel, 2
Sparus aurata, 208

Sphyraena obtusata, 209
Spinosum, 47, 53
Spiny lobster, 90
Spiny red weed, 51, 55
Spiral babylon, 113
Sponge, 174
Spot croaker, 190
Spotted seatrout, 189
Spotted sicklefish, 163
Spout, 262
Squalus acanthias, 256
Squeteague, 198
Squilla sp., 239
St. Peter's mark, 228
Starry triggerfish, 240
Steelhead (rainbow)trout, 225
Steller's sea lion, 261
Stenotomus caprinus, 207
Stolephorus commersonii, 149
Stomatopods, 171
Stomolophus meleagris, 72
Strap brown seaweed, 33
Streaked rabbit fish, 204
Striped bass, 177
Striped sea catfish, 237
Stromateus brasiliensis,
        209, 247
Strombus gigas, 112
*Strongylocentrotus
        droebachiensis*, 129
*Strongylocentrotus
        franciscanus*, 130
Stypopodium schimperi, 39
Sugar kelp, 42
Surf clam, 108
Sweet sea potato, 131
Swordfish, 213

T

Tadpole codling, 232
Tanner crab, 90
Tellina, 103
Tenualosa macrura, 145
Thalassoma fuscum, 163
Thalassoma trilobatum, 163
Thelenota ananas, 131
Theragra chalcogramma, 229
Thongweed, 24
Threadfin bream, 179
Thumbprint threadfin
        bream, 180
Thunnus alalunga, 196
Thunnus albacores, 195
Thunnus thynnus, 194
Tiger shark, 248

Todarodes pacificus, 120
Trachinotus blochii, 151
Trichiurus haumela, 212
Trichiurus lepturus, 212
Tripneustes gratilla, 128
Trumpeter whiting, 206
Tunicates, 161
Turbinaria conoides, 29
Turbot, 223
Two-spot red snapper, 175
*Tylosurus crocodilus
          crocodilus*, 234

U

Ulva clathrata, 17
Ulva faciata, 19
Ulva intestinalis, 18
Ulva lactuca, 3, 20
Ulva pertusa, 3, 17
Ulva reticulata, 21
Ulva spp., 3

Undaria pinnatifida, 43
Upside-down jellyfish, 40
Uroteuthis chinensis, 122

V

Valamugil seheli, 236
Velvet horn, 11
Venerupis decussata, 106
Venerupis philippinarum, 104
Viviparous, 242, 250
Vomerine teeth, 178, 241

W

Wakame, 43
Walrus, 265
Washington clam, 105
Weakfish, 189
White leg shrimp, 78
White sea bream, 208
White weakfish, 188
Wolf herring, 141

X

Xiphias gladius, 213

Y

Yellow spot goat fish, 178
Yellowfin tuna, 195
Yellownose skate, 252
Yellowtail amberjack, 153
Yellowtail scad, 159
Yellowtail, 161
Yesso scallop, 111

Z

Zeaxanthin, 8
Zooplankton, 89

Printed and bound by CPI Group (UK) Ltd, Croydon, CR0 4YY

21/10/2024

01777084-0009